Disclaimer

The publisher of this book is by no way associated with the National Institute of Standards and Technology (NIST). The NIST did not publish this book. It was published by 50 page publications under the public domain license.

50 Page Publications.

Book Title: Energy Consumption of Automatic Ice Makers Installed in Domestic Refrigerators (NIST TN 1697)

Book Author: David A. Yashar; Ki-Jung Park;

Book Abstract: This study examines the energy consumption of automatic ice makers installed in domestic refrigerators. The purpose of this research is to determine a method of measuring the energy consumption of automatic ice makers that will generate a repeatable and reproducible result. This study examined four refrigerator-freezers of different configurations, one top-mount unit, one side-by-side unit, and two French-door units with bottom freezers. The components and operational modes of each ice maker varied among the test subjects. This study examined each automatic ice maker and its components, and discussed how the operation of each component contributed to the overall energy consumption. Two of the units had a single-speed compressor that maintained the compartments within a range of temperatures by switching on and off. Since this type of unit draws considerable power when the compressor operates and very little when it doesn‰ot, a good representation for energy consumption can only be characterized over a number of whole compressor cycles. However, the unit‰os ice maker also produces cyclic variation in power draw; therefore accurate measurement of the ice maker energy consumption must be characterized over a number whole of ice making cycles. Since the unit‰os compressor cycles and the ice making cycles are not coincident, it is difficult to determine the ice making energy using a predefined test period. The other two units employed inverter driven variable-speed compressors, which operated by maintaining a constant, part-load condition and did not cycle on and off. Determination of the ice making efficiency for these units was rather straightforward since data could be examined over a whole number of ice making cycles. We also examined the conditions which affect the ice making efficiency.

Citation: NIST TN - 1697

Keywords: Refrigerator; ice maker;; energy consumption

NIST Technical Note 1697

Energy Consumption of Automatic Ice Makers Installed in Domestic Refrigerators

David A. Yashar
Ki-Jung Park

NIST National Institute of Standards and Technology • U.S. Department of Commerce

NIST Technical Note 1697

Energy Consumption of Automatic Ice Makers Installed in Domestic Refrigerators

David A. Yashar
Ki-Jung Park
Building Environment Division
Engineering Laboratory
National Institute of Standards and Technology
Gaithersburg, MD 20899-7321

April 2011

U.S. Department of Commerce
Gary Locke, Secretary

National Institute of Standards and Technology
Patrick D. Gallagher, Director

ENERGY CONSUMPTION OF AUTOMATIC ICE MAKERS INSTALLED IN DOMESTIC REFRIGERATORS

David A. Yashar and Ki-Jung Park

National Institute of Standards and Technology

Gaithersburg, MD 20899

Abstract

This study examines the energy consumption of automatic ice makers installed in domestic refrigerators. The purpose of this research is to determine a method of measuring the energy consumption of automatic ice makers that will generate a repeatable and reproducible result. This study examined four refrigerator-freezers of different configurations, one top-mount unit, one side-by-side unit, and two French-door units with bottom freezers. The components and operational modes of each ice maker varied among the test subjects. This study examined each automatic ice maker and its components, and discussed how the operation of each component contributed to the overall energy consumption.

Two of the units had a single-speed compressor that maintained the compartments within a range of temperatures by switching on and off (cycling). Since this type of unit draws considerable power when the compressor operates and very little when it doesn't, a good representation for energy consumption can only be obtained over a number of complete compressor cycles. However, the unit's ice maker also produces cyclic variation in power draw; therefore accurate measurement of the ice maker energy consumption should also be obtained over a number of complete ice making cycles. Since the units' compressor cycles and the ice making cycles are not coincident, it is difficult to determine the ice making energy using a predefined test period.

The other two units employed inverter-driven, variable-speed compressors, which operated by maintaining a constant, part-load condition and did not cycle on and off.

Determination of the ice making energy consumption for these units was rather straightforward since data could be examined over several complete ice making cycles.

We also examined the conditions which affect the ice making energy consumption. The results of these tests showed that the ice making energy consumption is influenced by the operating temperatures inside the cabinets. The results also showed that when their ice maker is switched from a mode where it is not producing ice to one where it is actively producing ice, some units exhibit temperature changes inside the cabinets without changing the thermostat settings. For these reasons, it may not be feasible to prescribe an ice making test at a specific thermostat setting, rather the results of multiple tests must be interpolated to a predefined set of compartment temperatures.

The measurements showed that only approximately one quarter of the energy used by a refrigerator to operate an automatic ice maker is actually used to freeze water into ice; the remaining three quarters is a result of using heaters to free the ice from the ice maker. The process of freeing ice from the molds may be accomplished by other, less energy intensive means; therefore, it is expected that introducing an ice making energy test will result in significant reductions of field energy use.

Our measurements indicate that the ice making energy consumption varies widely from product to product. The most efficient product tested consumed approximately 0.249 kWh per kilogram of ice produced under the test conditions used in this study, while the least efficient product consumed approximately 0.652 kWh per kilogram of ice. Variations in ice maker design and control algorithms played a large role in the ice production rate, which influenced the ice-making energy consumption. The annual energy consumption attributed to automatic ice makers added approximately 12 % - 20 % to the rated energy consumption value for each refrigerator, as determined by a regulatory test method. This value is based on the product of the measured ice making energy consumption per unit mass of ice produced and a consumer driven household ice consumption factor provided by a trade organization.

This study shows that it is possible to develop a test method for rating the energy use of automatic ice makers. In order to develop a test method that results in a repeatable and reproducible test result, the test must compare the energy consumed by a refrigerator with and without the ice maker being operated and the refrigerator maintaining comparable temperature conditions. Furthermore, data must be collected over a sufficiently long test period in order to reduce the influence of the refrigerator's compressor cycles operating at a different frequency from the ice maker's cycles.

Keywords: energy consumption, ice maker, refrigerator

Acknowledgement

This work was sponsored by the United States Department of Energy. Dr. W. Vance Payne contributed to the test apparatus design, operation, and data acquisition. Mr. John Wamsley provided technician support throughout various phases of this project. Ms. Natascha Milesi-Ferretti of NIST and Dr. Detlef Westphalen of Navigant Consulting International provided technical commentary on the draft of this report.

Table of Contents

List of Figures

List of Tables

Nomenclature

1: Introduction

Cyclic type automatic ice makers produce ice within one of the compartments of a domestic refrigerating appliance. They are directly connected to a source of water and continuously produce batches of ice and store it in a low temperature bin. At the present, these devices are rendered inoperative during the regulated energy consumption test (10 Code of Federal Regulations Part 430, Subpart B, Appendix A1, 2010) therefore the energy used to make ice is not measured. This is because ice making energy was not included in the basic energy consumption test (AHAM, 1979). The use of cyclic automatic ice makers do, however, have a significant impact on the product's energy use, which effectively goes undetected under current regulatory test procedure.

A cyclic type automatic ice maker consists of several components that enable repetitive cycling of three distinct sequential processes: (1) delivering of water into the ice maker, (2) freezing of water into ice, and (3) harvesting or removal of the ice from the molds in which it was frozen. The operation of these devices is halted when a sensor indicates that the ice storage bin is filled to its limit and can no longer accept new ice. The energy consumption associated with cyclic type automatic ice makers can be broken down into four categories. The first category is the energy used to deliver water into the ice maker; this is generally accomplished by energizing a solenoid valve. The second category is related to the additional load the refrigerator's vapor compression system must handle to extract thermal energy from the water in order to cool and freeze it. This second category of energy consumption is equal to the extracted thermal energy, the enthalpy difference between the water entering the system and the frozen ice, divided by the appliance's Coefficient of Performance (COP). The third category consists of the energy required to free the ice from the molds and deliver it to the storage bin. Most commonly, this is done by using electric resistance heaters to melt the interface between each ice cube and the mold where it was formed. The last category is related to the additional load to the appliance's vapor compression system needed to remove any thermal energy introduced to the cabinet through the other steps. Most commonly, this is the thermal energy introduced by ice heaters used to free the ice from the molds divided by the appliance's COP.

Limited research exists in literature on energy consumption measurement of cyclic automatic ice makers. Meier and Martinez (1996) attempted to quantify the energy consumed by automatic ice makers by producing approximately 500 g of ice and subtracting off the energy consumed under similar conditions with the ice maker inoperative. The approach used in this study, however, did not consider the variability of the mass of ice produced during the test, which was as much as 12 %, nor did it consider the temperature changes in the cabinets during ice maker operation. Another study by Haider et al. (1996) showed a more comprehensive evaluation of ice maker energy consumption. In this study, the researchers compared the energy consumption with and without the ice maker operating under two temperature settings and normalized the ice maker energy to the measured quantity of ice produced during the test. Although these authors did acknowledge substantial differences in compartment temperatures when the ice maker was rendered operative while using similar thermostat settings, and substantial differences in ice production rates with different thermostat settings, they did not interpolate the results of their measurements to a common condition.

These studies, however, were quite useful and have shown that the energy use associated with cyclic automatic ice makers is quite substantial compared to the total energy used by a domestic refrigerating appliance. These studies also suggested that energy used to cool and freeze water into ice does not constitute the majority of their energy consumption. Both studies show that the energy required to cool and freeze water into ice is approximately half of the total energy consumption of cyclic automatic ice makers; however, this proportion may be outdated. Continuous regulation of refrigerators have resulted large reductions in refrigerator energy consumption over the past 15 years, therefore the COP of modern refrigerators should be substantially higher than the products used in those studies. Both of these studies assumed that the test subjects had a COP of approximately 1.0. A higher COP reduces the component of total ice maker energy consumption related to the cooling and freezing of water, therefore making the impact related to the ice freeing process far more significant.

Considerations for Developing an Automatic Ice Maker Energy Test

Theoretically, it should be relatively simple to extract the energy used by a cyclic automatic ice maker simply by performing the basic energy consumption test with and without the ice maker operating, and examining the difference of the measurement results. In practice, however, there are several issues that complicate the development of a repeatable test. Some of the specific points are outlined below.

1) Ice-making cycles are not typically coincident with the appliance's compressor cycles. Most domestic refrigerator-freezers use a single-speed compressor that operates intermittently to maintain the compartment(s) within a range of temperatures. In order to accurately measure the energy consumption for a unit using this type of compressor, the power must be averaged over a whole number of complete 'on' and 'off' compressor cycles during an energy consumption test. Attempting to quantify the energy consumption with anything short of a whole number of compressor cycles introduces systematic error. Each ice-making cycle produces a discrete amount of ice, which is not obtained until the harvest step. Therefore, attempting to characterize the energy consumption of a cyclic automatic ice maker with anything short of a whole number of ice-making cycles will also introduce error. Ice-making cycles are generally independent of the compressor cycles; therefore any method used to measure an automatic ice maker's energy consumption must compromise between these two constraints. The error caused by this tradeoff can be minimized by extending the test period, which increases the number of whole compressor and ice making cycles under consideration.

   Furthermore, heat is removed from water at varying rates during the ice making process. Most cyclic automatic ice makers operate in such a way that water is frozen by cooling it with cold air. During the compressor 'on' time periods, the cold air near the ice molds is usually being circulated by fans; therefore the heat transfer mode used to remove thermal energy is forced convection. The heat transfer coefficients are roughly two orders of magnitude lower during the compressor 'off' periods because the fans are not running and heat is transferred to the surrounding air via free convection and conduction. Also, the air temperature oscillates during the compressor cycles (decreasing temperatures during the on cycles and increasing temperatures during the off cycles) and the temperature of the water in the molds is continuously decreasing as it approaches a frozen state. Therefore, as the water is cooled and frozen into ice, the heat transfer rate governing the production of ice varies widely because of the variations in heat transfer coefficients, water temperature, and air temperature. This aspect has an influence on the repeatability of an ice-making energy measurement and it compounds the effect of the non-coincident compressor and ice-making cycles because it is dependent on the temporal phase shift between the ice-making cycles and the compressor cycles.

2) <u>Cabinet temperatures may depart from steady-state operation in response to ice-maker operation.</u> When an automatic ice maker inside an appliance is rendered operative, one or more of the appliance's compartments may exhibit some type of temperature change. This response varies by product configuration and control algorithm. Certain units respond in such a way that water introduced into the ice maker's mold in one compartment results in a much higher temperature in that compartment due to the thermal load of the warm unfrozen water. Other units respond in such a way that they overcompensate for this load and actually cool this compartment below the temperature that was maintained before ice making was initiated. Another variation is that the cabinet's control system will simply have difficulty balancing the load between the fresh food and freezer compartments when making ice, therefore inducing a temporary departure from the steady-state temperature of both the fresh food and frozen compartments. It should be noted that the most highly sophisticated state-of-the-art units with electronic controls generally don't exhibit temperature changes of the same magnitude as the traditional mechanically-controlled units. This aspect raises the question of whether or not to consider cabinet temperature departures during ice making. This can be done by interpolating the results of multiple ice-making tests to calculate the ice making capacity and energy consumption at a set of standard temperatures. This could then be a uniform basis for comparison with the basic energy test. The aspect of temperature departure should be considered in some way; however, there are at least two reasonable approaches to consider this aspect. One method is to perform multiple ice making tests and interpolate the results of these tests to a given target temperature. Another method is to perform a single ice making energy consumption measurement and use the recorded temperatures from that data as a target set of temperatures for interpolation with the baseline energy consumption measurements (without making ice). Therefore this could consist of a single ice-making measurement in addition to the already existing energy consumption measurements.

3) Determination of an appropriate start and end point for an ice-making test. As mentioned earlier, the ice making cycles consist of (a) filling the molds, (b) freezing the water into ice, and (c) ejecting the cubes. Most, if not all, ice makers fill the molds shortly after ejecting the cubes from the previous cycle. When an ice maker is rendered inoperative, usually by sensing a full storage bin, it halts the cycle by ceasing the ejection operation; therefore, ice is held in the molds until the storage bin condition is no longer full. There are several reasons why this is desirable from an end user's point of view. However, it complicates the ice-making energy characterization because engaging the ice maker will reinstate the ice-making cycle just prior to the ejection step, thereby negating the energy associated with filling the molds and freezing the water for the first batch of ice produced. This issue was addressed by the studies of Meier and Martinez (1996) and by Haider et al. (1996), by using a valve upstream of the water connection to the refrigerator in order to control the flow to the ice maker. This allowed these researchers to have some control over the beginning and end of the ice maker operation.

4) Determination of an appropriate inlet water temperature. The temperature of water supplied to the cabinet has an impact on the overall ice-making energy consumption because it is directly related to the amount of energy that must be extracted to freeze it into ice. In a typical household installation, water is supplied to the cabinet via direct connection to a water source generally maintained between 10 °C and 21 °C (50 °F and 70 °F). However, water in the supply lines to the cabinet moves very slowly which causes the water to reach thermal equilibrium with the environment before entering the cabinet. Therefore, attempting to supply water to the cabinet at any temperature other than the ambient temperature in the test chamber is not feasible. The ambient room temperature during the current energy consumption tests is 32.2 °C (90.0 °F). Using 32.2 °C (90.0 °F) as the temperature of the water entering the ice maker is an atypical condition and may present a biased challenge to a unit under test; however, using a temperature more representative of field conditions would result in greater test burden.

5) Determination of how to treat interactions with defrost cycles. Most domestic refrigerators periodically undergo a cycle that removes accumulated frost from the evaporator. During this period of operation, the power consumption and cabinet temperatures undergo deviations from steady state operation. When a defrost cycle occurs during a time when an ice maker is rendered operational, analyzing the energy consumption measurement data can get quite complicated because these are two transient phenomena working against each other within the cabinet. Without going into a great level of detail, it is undesirable to use data that encompasses any portion of a defrost sequence when examining the ice making energy. This is a complication because the test operator generally has no control over the timing of a defrost cycle and cannot easily structure data collection to avoid defrost periods. Furthermore, in many cases the production of ice shortens the time between successive defrost sequences because many refrigerators employ

4

some type of demand defrost and the supply of water into the ice maker often generates an accelerated demand for defrosting.

6) <u>Non-standard ice production rate</u>. Every model of automatic ice maker produces ice at its own rate, therefore it may produce a biased evaluation to rate such a device on the basis of energy consumed per unit of time. The more appropriate approach is to rate them by ice-making energy per unit mass of ice produced. This metric can later be converted to energy/time based on a standardized usage factor of mass of ice per unit time. With this in mind, it is important to note that while testing in a 32.2 °C (90 °F) ambient would reduce the test burden, the amount of energy that must be removed from the water to freeze it is greater than what would normally be expected in field operation. Therefore, the ice production rate under test conditions would likely be smaller than in the field.

Because of all of these known difficulties, there have also been suggestions that a purely calculation based approach for determining ice maker energy use might yield a better result. The suggested approach is to tabulate the energy used by each of the cyclic automatic ice maker's components and add that to the additional compressor work which is the total thermal load (energy removal from water/ice plus the load from any heaters used) divided by the system's COP. This approach will also be investigated as an option to gauge ice maker energy consumption.

This study examines the ice making energy associated with cyclic automatic ice makers installed in four different styles of domestic refrigerator-freezers. The objective of this study is to facilitate the establishment of a robust, repeatable ice making energy consumption measurement method that would be applicable to cyclic automatic ice makers and would not substantially increase the test burden beyond that of the currently regulated energy consumption test. This study characterizes the energy consumption attributed to four automatic cyclic ice makers by examining the difference between the energy consumption of each refrigerator-freezer with the ice maker operative and inoperative. Various interpolation methods are examined and their merits and drawbacks are compared. Specifically, we examined the energy consumption of an automatic ice maker by (1) examining the ice-making and non ice-making energy consumption of a unit at a single set of temperature settings; (2) examining the ice-making and non ice-making energy consumption of a unit using multiple measurements and interpolating the results to a common set of operational conditions, this was performed using two different interpolation methods; and (3) a purely calculation based approach.

During the data analysis, the individual data sets will be examined to determine an appropriate length of time for data collection, since steady state operation does not generally exist.

2: Test Setup and Data Acquisition

The test setup was constructed in accordance with the Department of Energy's test procedure outlined in 10 Code of Federal Regulations Part 430, Subpart B, Appendix A1, 2010. There was one slight modification to this setup; a continuous water supply was connected to the unit under test to supply water to the automatic ice maker. Two identical test cells were constructed to simultaneously measure the energy consumption of two refrigerators. Each test cell consists of a thermally non-conductive platform and a vertical wall adjacent to one edge of the platform. Each refrigerator is mounted on top of the platform with its rear adjacent to the vertical wall. All faces of the test cells were painted dull black to minimize the radiant heat transfer to and from each refrigerator during testing.

The test cells were placed in an environmental chamber, which was large enough to house two test cells and all of the necessary data acquisition hardware. This environmental chamber was capable of providing controlled ambient temperature and humidity over long periods of time with little supervision, as necessitated by the lengthy test periods of domestic refrigerator energy consumption measurements.

Water supply lines were connected to each test subject. The temperature of the water supplied to each unit was not controlled, but the tube connecting each line to a refrigerator was sufficiently long to allow the water to equilibrate with the temperature in the chamber. This was verified by inserting two T-type thermocouples into each water line just upstream of the refrigerator connection.

All of the temperature and humidity data were gathered using a personal computer and a multiplexed data acquisition unit. The electrical energy input was monitored using a separate personal computer dedicated to two digital power meters, one connected to each test unit. All temperatures were sampled every 30 s, and the power was sampled every two seconds. Table 2.1 lists the measured quantities and the uncertainty associated with 95 % confidence. The equations used to calculate the measurement uncertainty are shown in the Appendix.

Table 2.1: Measurement Uncertainty

Measured Quantity	Measurement Device	Uncertainty at 95 % confidence
Temperature	Thermocouples	± 0.1 °C (0.2 °F)
Power	Watt-meter	± 0.5 % of reading
Energy	Watt-meter	± 0.5 % of reading
Mass	Digital Scale	± 5 gram

The dry-bulb temperature in the environmental chamber was maintained at 32.2 °C (90.0 °F) in accordance with the US DOE test procedure. Throughout the duration of the tests, the analysis was performed to characterize the performance of the test subjects while maintaining cabinet temperatures of 3.9 °C (38 °F) in the refrigerator compartment and -17.8 °C (0 °F) in the freezer compartment. These target temperatures are different from the current Department of Energy test procedure, 7.2 °C (45 °F) in the refrigerator compartment and -15 °C (5 °F) in the freezer compartment, but will be incorporated in the upcoming test procedure, therefore it is of greater interest to use these new temperatures. It is important to note, however, that the test subjects in this study were rated using the current target temperatures, therefore it is expected that they will not perform exactly to their rated specification.

In total, four domestic refrigerator-freezers were used for this study. The units were selected to provide information spanning a variety of designs and features that are expected to influence ice making energy consumption. The first unit in this study is a very typical model, a top-mount refrigerator-freezer with an ice maker located in the freezer compartment. The second unit is a side-by-side refrigerator-freezer with an automatic ice maker located inside the freezer door, and it dispenses ice through the freezer door. The third and fourth units are French door style units, whose freezer compartments are located beneath their refrigerator compartments. The third unit produces ice by directing some cold air from the freezer to an insulated ice-making sub-compartment in the refrigerator compartment. Lastly, the fourth unit produces ice by means of a separate ice making evaporator in an insulated ice making sub-compartment located in the refrigerator compartment

3: Experimental Results for Top Mount Refrigerator Freezer

The first test unit is a 600 liter (21 cubic foot) top mount, energy star rated model. It has variable defrost control, mechanical thermostat control, and a single-speed compressor. The unit has a factory installed cyclic automatic ice maker, which was connected to an external water supply and produces and stores ice in a bin located in the freezer compartment. The unit was outfitted with three thermocouples in the refrigerator compartment and three thermocouples in the freezer compartment, as shown below. The compartment temperatures used for the calculations that follow were the average of the time averaged values reported from the three thermocouples located in each compartment during the test duration.

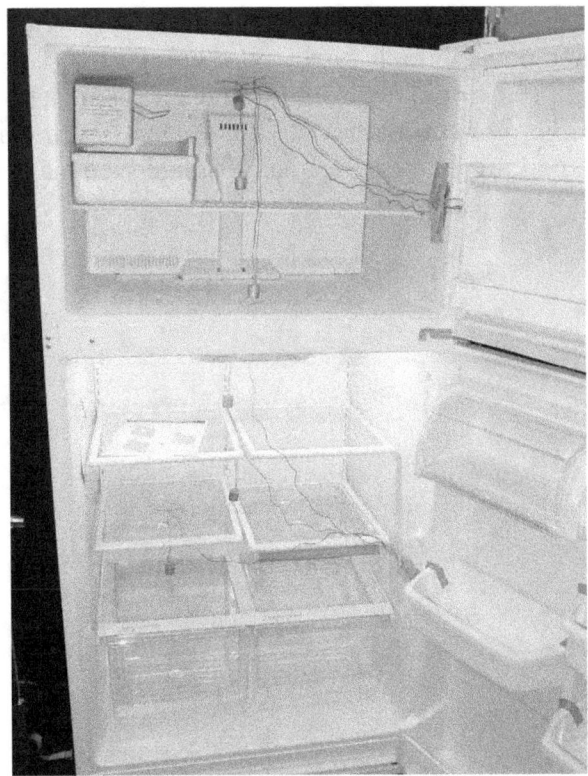

Figure 3.1 Top Mount Refrigerator Freezer

The minimum and maximum time between defrost cycles for this unit is 8 hours and 96 hours of compressor run time, respectively. According to the procedure outlined in the current DOE energy test procedure, these values result in a calculated average time between defrost of 30 hours of compressor run time.

3.1 Top Mount Refrigerator Freezer with No Ice Production
3.1.1 Mid Setting Results
During the first test, the energy consumption was measured with the thermostats set to the median setting for each compartment. The following results were obtained:

Steady state cyclic operation time: 12456 seconds = 03:27:36

Measured refrigerator compartment temperature: (3.2 ± 0.1) °C
Measured freezer compartment temperature: (-19.3 ± 0.1) °C
Energy expended during the test period: (190.9 ± 0.9) watt-hours

Alternatively, this can be expressed as an average steady state power of (55.18 ± 0.28) watts at the measured temperatures.

The defrost sequence for this unit consisted of (a) defrost heater switching on, (b) defrost heater switching off, (c) period of no power draw, (d) compressor on – recovery cycle, and (e) the following compressor off cycle. Measurements during this sequence yielded the following results:

Defrost time: 6151 seconds = 01:42:31
Energy expended during defrost period: (209.6 ± 1.0) watt-hours

Combining the results of these two portions of the test, assuming that the average time between defrosts is 30 hours of compressor run time, we arrive at the total energy consumption of (500.3 ± 2.5) kWh/yr at the measured temperatures.

3.1.2 Warm Setting Results
Since the temperatures in both compartments for the first test were colder than the target temperatures, the thermostat was set to the warmest setting and the measurements were performed again. The following results were obtained:

Steady state cyclic operation time: 11346 seconds = 03:09:06
Measured refrigerator compartment temperature: (6.0 ± 0.1) °C
Measured freezer compartment temperature: (-14.5 ± 0.1) °C
Energy expended during the test period: (147.3 ± 0.7) watt-hours

Alternatively, this can be expressed as an average steady state power of (46.73 ± 0.23) watts at the measured temperatures.

Defrost time: 4998 seconds = 01:23:18
Energy expended during defrost period: (174.0 ± 0.9) watt-hours

These values resulted in the total energy consumption of (425.3 ± 2.1) kWh/yr at the measured temperatures.

3.1.3 Combined Test Results for Top Mount Refrigerator Freezer with No Ice Production
We then combined the results of the measurements at the first thermostat setting with the results at the second thermostat setting and interpolated to the target temperatures. When interpolating to the refrigerator target temperature of 3.9 °C, the calculated energy consumption was (481.6 ± 2.3) kWh/yr; when interpolating to the freezer target temperature of -17.8 °C, the energy consumption was (476.9 ± 1.6) kWh/yr. The maximum of these two values is declared as the baseline energy consumption: (481.6 ± 2.3) kWh/yr.

It is also useful for later calculations to note the energy consumption obtained through interpolation if the defrost energy was not considered. In this case, the calculations were performed with the steady state operational average power of (55.2 ± 0.3) W $((483 \pm 2)$ kWh/yr) from the mid/mid setting and (46.7 ± 0.2) W $((409 \pm 2)$ kWh/yr) from the warm/warm setting rather than the total values which included the contribution of the defrost cycles. Here, negating the energy from periodic defrosting shows the energy consumption for this unit to be (465 ± 2) kWh/yr.

3.2 Top Mount Refrigerator Freezer with Ice Production
With the baseline energy consumption characterized, we next performed a series of measurements to examine the energy consumption attributed to the production of ice by the automatic ice maker.

3.2.1 Mid Setting Results
The thermostat was set to the median position for each compartment and the ice maker was rendered operative. Water was supplied to the unit at a temperature of 32.2 °C. The power signature of this unit showed a few differences compared to the measurements with the ice maker inoperative. Figure 3.2 below shows the power drawn by the unit with the thermostats set to the median setting with (a) ice maker inoperative and (b) ice maker operative.

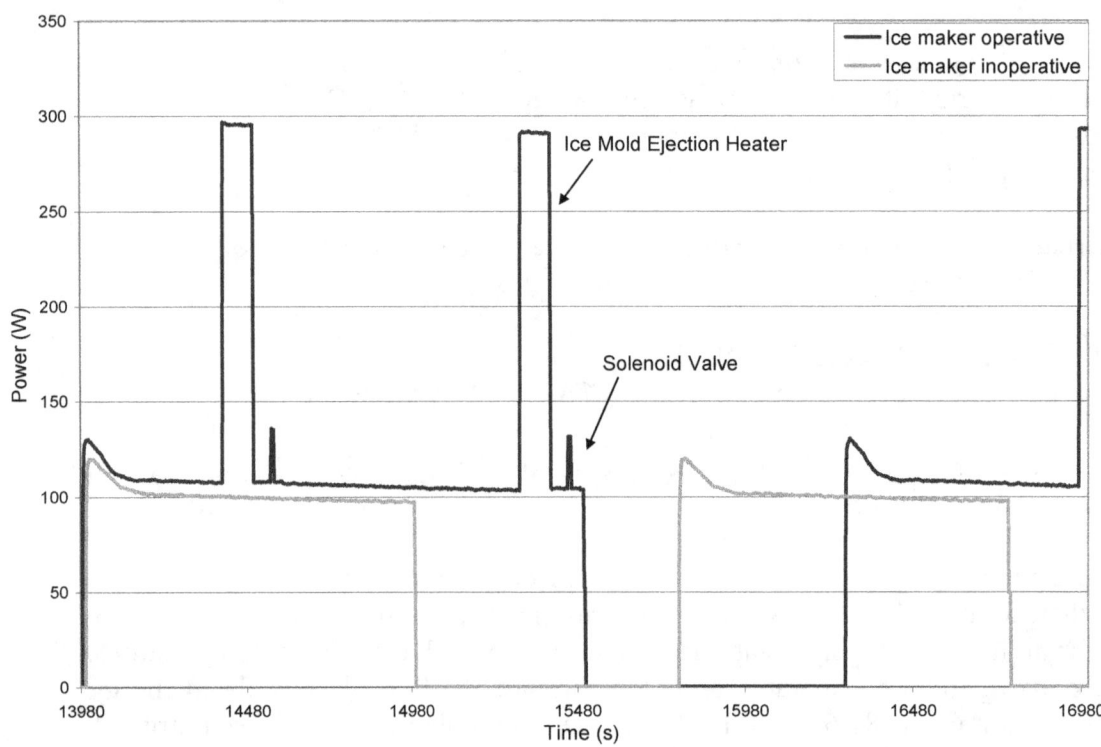

Figure 3.2 Power Signatures of Top Mount Unit under Test

10

The first difference between the signatures for the two different ice-maker settings is that the compressor on cycle is much longer while the unit is producing ice; this is due to the additional cooling load. The second difference is the power consumed by the ice-maker components. There are two separate energy-consuming actions that are engaged during the process of ice making and harvesting. The first block of energy is consumed by the ice mold ejection heater, which draws approximately 190 watts for a period of 90 seconds. This heater is used to heat up the tray in which the ice is formed so that it can be freed and dropped from the mold into the storage bin. The second device is a solenoid valve, which is powered up to allow water to flow from the supply into the ice molds; this solenoid draws approximately 28 watts for a period of 10 seconds. It is also very important to note that the ice making cycle, which consists of the time period starting from the solenoid valve fill through the operation of the ice ejection heater, is completely independent from the compressor cycle. Since the results of all measured quantities are averaged over a whole number of successive compressor cycles, it is desirable to record data over the longest attainable time period in order to increase the number of whole compressor cycles. Extending the data set in this manner increases the accuracy of the results by reducing the effects of the incomplete ice making cycles.

Another aspect of the ice-making test data was that the additional load to the freezer compartment resulted in an elevated temperature in this compartment. Figure 3.3 shows the temperatures measured during the ice-making test. At a certain time (approximately 60 000 seconds on the graph) the ice maker automatically halted ice production due to the ice storage bin reaching its fill limit. After this point the freezer compartment temperature decreased by approximately 3 K which resulted in a temperature similar to that seen during the test at the same thermostat setting test with the ice maker inoperative.

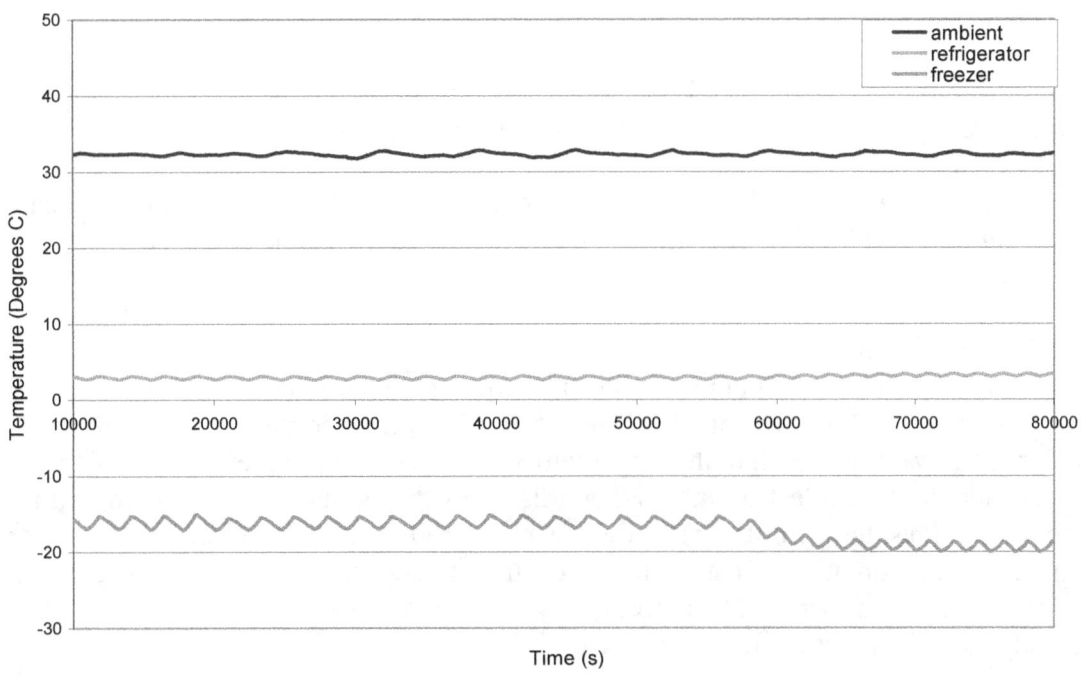

Figure 3.3 Temperature Response to Ice Maker Operation

11

The data recorded during the time period encompassing 25 whole compressor cycles and all ice production showed the following results:

Steady state cyclic operation time: 57390 seconds = 15:56:30
Measured refrigerator compartment temperature: (2.9 ± 0.1) °C
Measured freezer compartment temperature: (-16.2 ± 0.1) °C
Mass of ice produced: (1.34 ± 0.005) kg
Ice production rate: (84.1 ± 0.3) grams per hour
Energy expended during the test period: (1.355 ± 0.007) kWh

Alternatively, this can be expressed as an average steady state power of (84.99 ± 0.43) watts at the measured temperatures.

Single Data Point Method for Measuring Ice-Maker Energy Consumption
The first proposed approach to quantifying the ice-making energy is to compare the steady state energy consumption with both thermostats set to their median setting for both ice producing and ice maker inoperative conditions. In this case, the differential power consumption was:

$$(84.99 \pm 0.43) \text{ W} - (55.18 \pm 0.28) \text{ W} = (29.81 \pm 0.51) \text{ W}$$

The energy used to produce ice in this configuration can then be calculated from the mass of ice, the differential power and the test period time.

$$\frac{(0.0298 \text{ kW})(15.941 \text{ h})}{(1.34 \text{ kg})} = (0.355 \pm 0.006) \text{ kWh/kg}$$

The uncertainty of ± 0.006 kWh/kg was calculated using the individual uncertainties of the measured quantities. This notation is used throughout this report.

This, however, may not be the best approach to estimate the ice-making energy because the temperatures realized in this unit changed considerably in response to the activation of the ice maker.

3.2.2 Cold Setting Results
The next approach was to attempt to use an interpolated energy (and ice production rate) using a second set of measurements. Since the internal temperature of the freezer compartment was warmer than the target temperature, we set the thermostats of each compartment to the coldest setting. When this unit was operated with the compartments set to their coldest temperatures and warm water introduced into the freezer through the ice maker, the automatic defrost control algorithm made the unit operate with a very short time between defrost periods. The frequency of these defrost periods limited the amount of steady state data that could be obtained.

The power measurements are shown in Figure 3.4. During these tests, the ice maker was activated at the beginning of a defrost cycle in order to maximize the amount of useable data. The operation of the ice mold ejection heater is clearly visible in this figure, and its deployment can be used as a counter for the number of ice making cycles during the test.

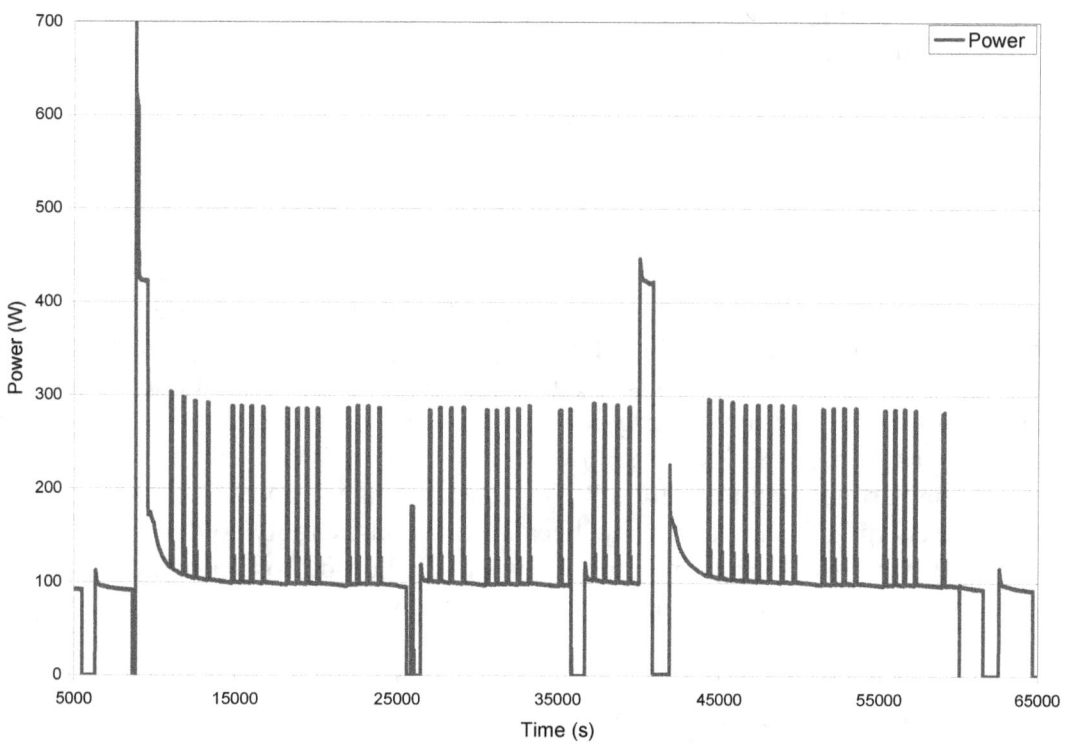

Figure 3.4 Refrigerator Power at Cold Setting with Ice Production

The ice mold ejection heater is first energized during the same time period as the defrost heater. At the termination of the defrost heater operation, the compressor turns on and undergoes a long on period, called the recovery period. The ice ejection heater is energized 16 times during the recovery period and one more time during the compressor off cycle following the recovery cycle. Next, there is a single "normal" compressor cycle, during which time the ice ejection heater is energized 11 times. The compressor cycle following this "normal" cycle leads into the next defrost sequence, during which time the ice-ejection heater is energized 4 more times. The data from this set of measurements must be used to estimate the marginal energy consumption due to ice making; therefore we have to consider only the steady state operation portion, i.e. that which in not influenced by either defrost events. We need to calculate the relevant values averaged over a whole number of compressor cycles operating at steady state; however, there simply aren't enough compressor cycles to satisfy the steady state criteria set forth in the test method. Therefore the only reasonable approach is to use the data from the single "normal" compressor on/off cycle.

Single cycle operation time: 9758 seconds = 02:42:38
Measured refrigerator compartment temperature: (-2.5 ± 0.1) °C
Measured freezer compartment temperature: (-20.0 ± 0.1) °C

Mass of ice produced: (0.315 ± 0.005) kg
(estimated by multiplying the total mass of ice produced by 11/50 – the proportion of ice ejection heater cycles that occurred during the normal compressor cycle)
Ice production rate: (116.2 ± 1.8) grams per hour
Energy expended during the test period: (301.6 ± 1.5) watt-hours
Average steady state power of (111.28 ± 0.56) watts

Two Data Point Method for Measuring Ice-Maker Energy Consumption
Now, the energy consumption at the specified target temperatures is calculated based on the results of the mid/mid and cold/cold setting tests. For this calculation, the interpolated values for a freezer temperature of -17.8 °C are an energy consumption of (841.5 ± 5.0) kWh/yr. Since the ice production rate varied between these two tests, it was necessary to estimate the ice production at the target temperatures by some means. Therefore, the ice production rate at the target temperatures was estimated using a linear interpolation approach, this effectively weighted the measured ice production rates in the same proportion as the measured energy. The ice production rate at the target conditions was calculated as (855.1 ± 8.6) kg/yr. However, these interpolated values occur at a refrigerator compartment temperature of 0.74 °C, which manifests itself as a substantial penalty on the energy consumption value because it is considerably colder than the target temperature. Regardless, comparing these values to the non-defrost influenced data for the tests without ice production results in the following values:

Energy consumption difference:
(841.5 ± 5.0) kWh/yr - (464.5 ± 2.2) kWh/yr = (377 ± 5.5) kWh/yr
Ice production rate: (855.1 ± 8.6) kg/yr
Ice production energy: (0.441 ± 0.008) kWh/kg

It should again be noted that this unit has difficulty controlling the internal compartment temperatures during the ice production phase, and that operating this unit at the same thermostat settings with and without ice production does not result in the same compartment temperatures. This is because ice production places a large processing load onto the duty of the freezer compartment without any addition to the refrigerator compartment's duty. Also, the mechanical thermostat used by this unit has difficulty balancing the cooling load requirements between the compartments in ice production mode.

3.2.3 Mixed Setting Results
In order to attempt to obtain a more accurate representation of the ice-making energy, another data set was taken with the freezer compartment set at the coldest setting and the refrigerator compartment set at its warmest setting. By using three data points, a more accurate triangular interpolation method (ASNZ, 2007) can be applied in lieu of the linear interpolation method. The following values were obtained from this data set.

Steady state cyclic operation time: 16805 seconds = 04:40:05
Measured refrigerator compartment temperature: (3.16 ± 0.1) °C

14

Measured freezer compartment temperature: (-17.1 ± 0.1) °C

Mass of ice produced: (0.419 ± 0.005) kg

  (estimated by multiplying the total mass of ice produced by 15/48 – the proportion of ice ejection heater cycles that occurred during the normal compressor cycles)

Ice production rate: $(89.7 \pm 1.1))$ grams per hour

Energy expended during the test period: (399.5 ± 2.0) watt-hours

Average steady state power: (85.58 ± 0.43) watts

Three Data Point Method for Measuring Ice Maker Energy Consumption

The triangular interpolation method is based on the idea that one can estimate the energy consumption of a two compartment cabinet at a specific pair of internal temperatures by using three data points that form a triangle around that pair of temperatures. Using the three measured data points, the position of a theoretical fourth data point (T4) is calculated, which lies at the intersection of a line joining two of the data points with a line joining the third data point to the target set of temperatures. This theoretical data point serves as a basis to calculate the energy consumption at the target set of temperatures. Figure 3.5 below shows the data points in graphical form. For this particular unit, however, the mechanical thermostat had limited ability to control the compartment temperatures during ice production. Therefore, we were unable to acquire a third data set that graphically forms a triangle with the other data sets that contains the set of target temperatures of 3.9 °C and -17.8 °C; this is why the target temperatures lie outside of the triangle formed by the three measured data sets. The mathematics outlined in this method can still be applied to extrapolate the data to the target temperatures. It should be noted that the distance between the data set from the coldest setting is far from the other two data points and the target point, which causes the calculated measurement uncertainty to be large compared to the other data sets in this study.

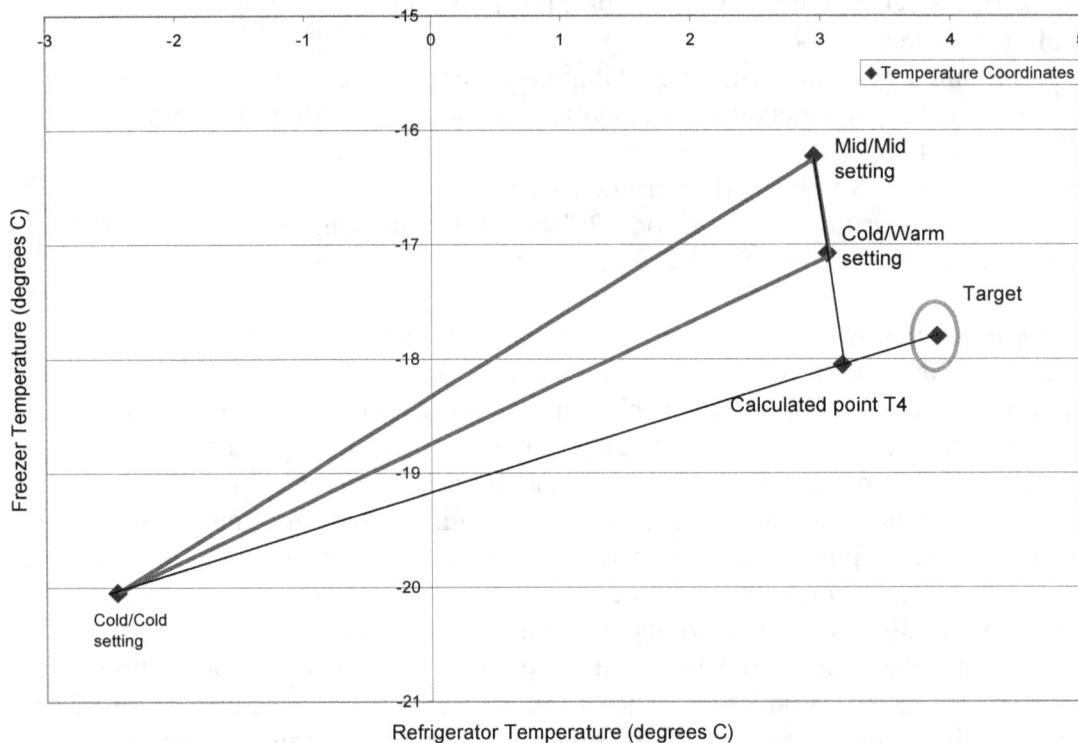

Figure 3.5 Graphical Representation of Triangular Interpolation

Using this approach, values for the energy consumption and ice production rates were calculated as if they were obtained at the target temperature conditions.
Energy consumption: (732 ± 16) kWh/yr
Ice production rate: (816 ± 87) kg/yr

Comparing this to the baseline energy consumption value without ice productions, the result is:

Energy consumption difference:
(732 ± 16) kWh/yr - (464.5 ± 2.2) kWh/yr = (267 ± 16) kWh/yr
Ice production energy: (0.327 ± 0.04) kWh/kg

All of the above methods estimate the amount of energy consumed for the purposes of making ice while excluding the interactions between the ice production and the defrost sequence. It is very difficult to separate out the ice-making energy if these interactions are considered; particularly since ice making may, but does not necessarily occur during the defrost sequence. Also, this particular unit shows that operating the ice maker influences the automatic defrost control scheme, which may not be the case for all refrigerator freezers, and it is therefore questionable whether this interaction could be neglected. Furthermore, since ice making is prompted by user interaction and automatically halted when the storage bin is filled, it is difficult to determine the significance of these interactions during field use. Obviously, ice production is a much

16

less efficient process during the time period that the defrost heater is operating, but this situation may have limited occurrence.

3.3 Calculation Based Method

The last method examined for determining ice-maker energy consumption is a simplified calculation approach. There are four components contributing to energy consumption during ice production: (1) energy used by the solenoid valve to deliver water to the ice molds, (2) the energy consumed to run the refrigeration cycle to meet the demand of removing heat from the water, (3) the energy used by the ice mold ejection heaters to free the ice from the molds, and (4) the energy used to run the refrigeration cycle to meet the demands of the additional load contribution of the ejection heaters.

Each data set taken with the ice maker operative was started with the ice bin empty and logged until the bin reached its shut off limit. The first data set showed 1.34 kg of ice produced through 47 cycles, the second showed 1.43 kg in 50 cycles, and the third showed 1.34 kg in 48 cycles. Therefore, this unit produces an average of 28.3 grams of ice during each cycle.

The solenoid valve draws 28 watts for a period of 10 seconds to deliver water to the molds, which results in 280 joules per ice-making cycle. The ice ejection heater draws 190 watt for a period of 90 seconds to free the ice from the molds, which results in 17,100 joules per ice making cycle. The total energy consumed by these components is therefore 17,380 joules per ice making cycle. Considering the mass of ice produced per cycle, this can be expressed as 614.1 kJ/kg of ice produced. Also, the energy consumed by the ice ejection heater is ultimately dissipated as heat into the freezer cabinet; therefore it must be considered an additional heat load to be processed by the refrigeration system.

The energy required to cool 32.2 °C water to freezing is comprised of three parts. First, approximately 135 kJ/kg of energy is required to reduce the water's temperature to 0 °C. Then 334 kJ/kg of energy is required to freeze it into ice. Lastly the ice must be cooled down to the freezer temperature (-17.8 °C) by expelling an additional 36 kJ/kg. In total, this accounts for 505 kJ of energy per kilogram of ice.

The proposed calculation method relies on an assumed value for a Coefficient of Performance (COP) of the unit. At any given set of operating conditions, modern products operate with a somewhat broad range of COP's and it is difficult to select a representative value that is applicable for all appliances. For the sake of exploring the calculation approach, we will use an assumed COP value of 1.8, i.e. that the system can remove 1.8 units of heat for every unit of electrical input. It should be emphasized that this assumed COP is merely a placeholder assumption and has no verifiable basis.

While making ice, this refrigeration system must meet the additional demand of 1119 kJ of heat per kg of ice produced (505 kJ for cooling and freezing the water plus 614.1 kJ for the load from the ejection heaters). Using the assumed value of the COP, the additional electrical input to the system required to meet this load is 621.7 kJ/kg of ice produced. Adding in the energy consumed by the solenoid and ejection heater brings the total load to 1236 kJ/kg; or in alternative units 0.343 kWh/kg.

3.4 Test Length Determination

The main point of this exercise is to examine the power response from the unit while making ice in a little more detail in order to determine how much data needs to be recorded to properly calculate the energy consumption. Figure 3.6 shows the average power drawn during each full compressor cycle while the unit was operating at the mid/mid setting and producing ice.

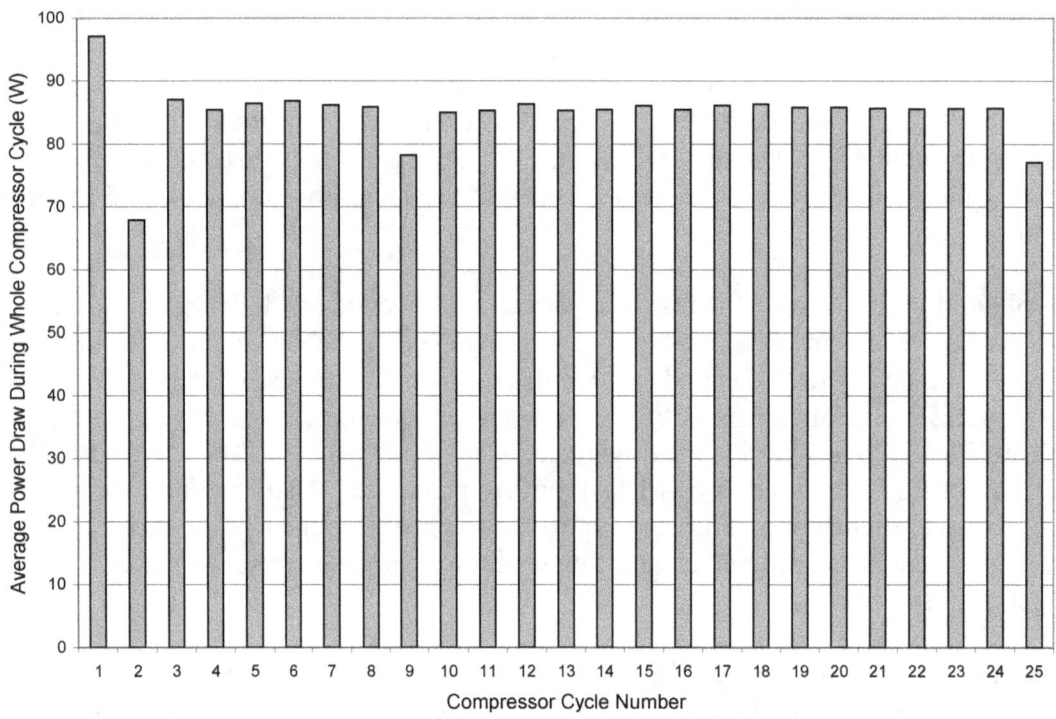

Figure 3.6 Average Power Draw Over Full Compressor Cycles – Top Mount Unit Test #1

Throughout this test period, there were generally two batches of ice harvested during each compressor cycle. The exceptions are cycle number 1 (3 batches), 2 (0 batches), 9 (1 batch), and 25 (1 batch). This is due to the fact that the ice-maker cycles are not perfectly coincident with the compressor cycles.

Figure 3.7 shows this same data plotted as the cumulative average power draw versus time; the instantaneous power is also shown in the figure. Here, the beginning of this data is coincident with the compressor switching on; therefore each local minimum point on the plot is indicative of another compressor on switch. Consequently, each local minimum also represents the cumulative average power draw over a whole number of compressor cycles. We can see that as time elapses, the local minima of cumulative average power draw asymptotically approaches a value of approximately 85 watts. This value represents the average power draw needed to maintain the temperatures in the cabinet and simultaneously produce ice over a sufficiently long test period, which is the value that is of interest to this study. It is interesting to note that the running average stays within 1 % of the final value after only four ice ejection cycles were recorded. This means that the average power could be determined in a test period consisting of as little as four compressor cycles for this unit under these test conditions and be within 1 % of the steady state value.

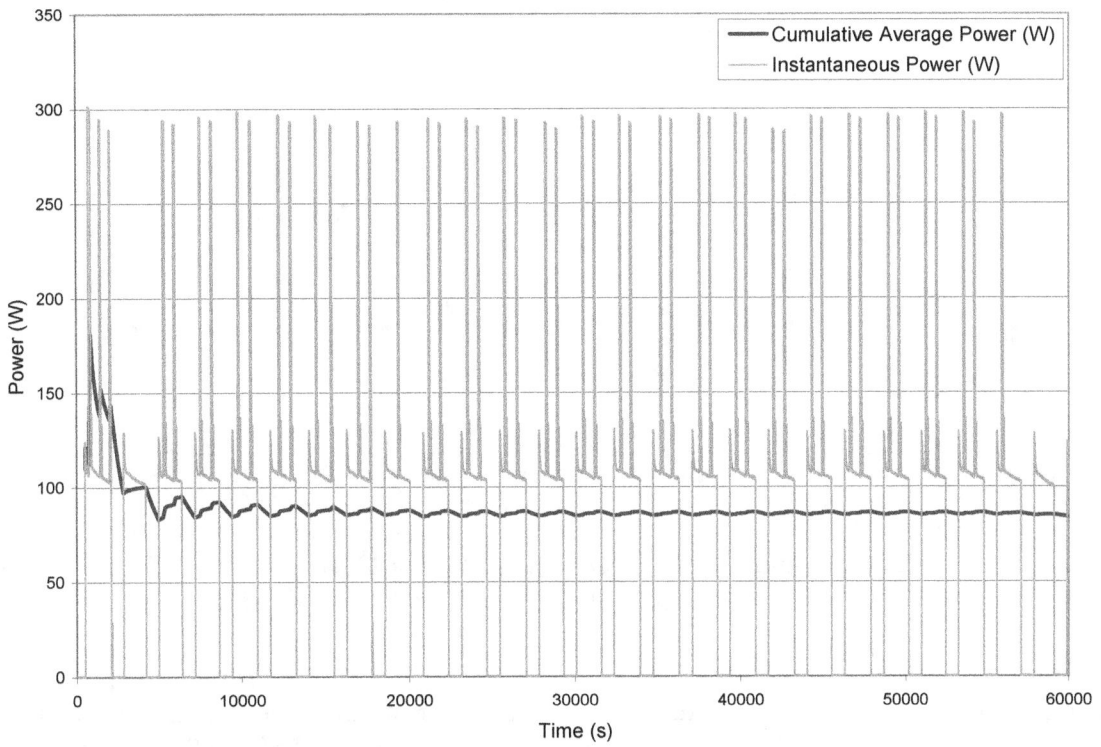

Figure 3.7 Instantaneous and Cumulative Average Power Draw
Top Mount Unit Test #1

Figure 3.7 shows that while the cumulative average power is stable after 4 compressor cycles, the ice production rate is not as stable. During the first 4 compressor cycles, the unit produced 7 batches of ice, alternatively 1.75 batches of ice per compressor cycle; while the full data set shows that it produced 47 batches of ice in 25 compressor cycles, or 1.88 batches per compressor cycle. Since the ultimate goal is to determine the test length required to stabilize the value for the energy required to produce ice, the criteria should therefore be based on the cumulative energy normalized to the amount of ice produced, shown in Figure 3.8. According to this approach, the ratio of cumulative energy per batch of ice stabilizes within 1 % of the final value after 17 compressor cycles, or approximately 10 hours of test time. The data shown in the figure also illustrates that 2 % stability is reached in 14 compressor cycles (9 hours) and 4% is reached in 10 cycles.

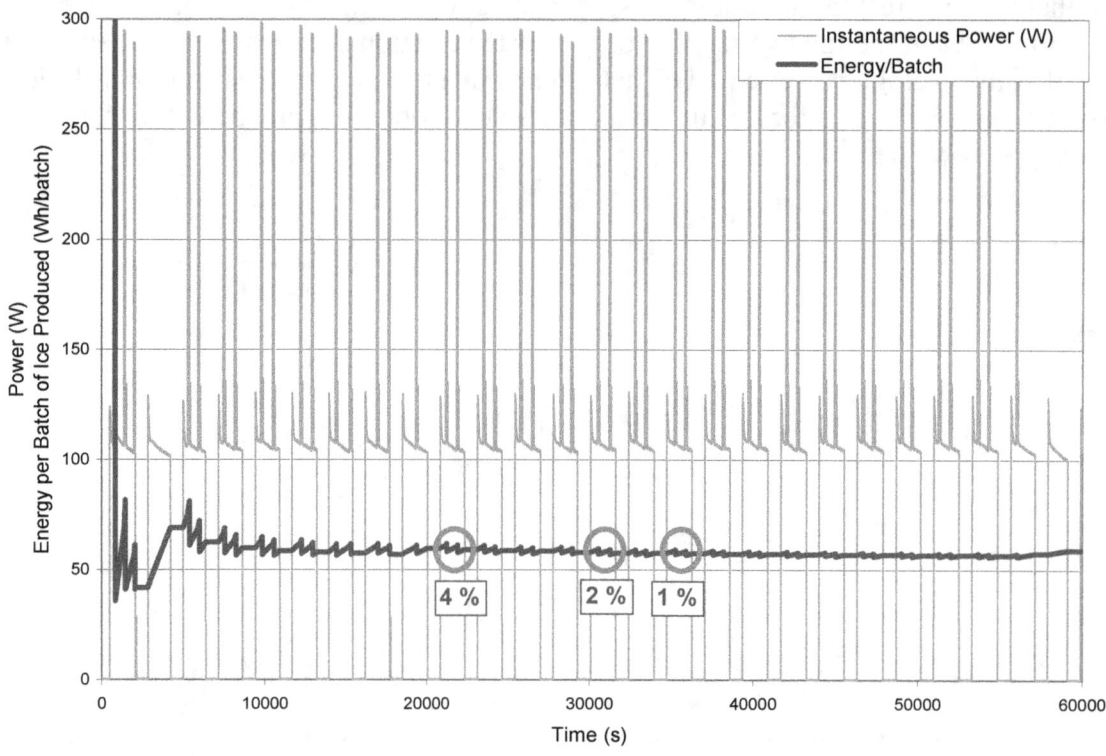

Figure 3.8 Instantaneous and Cumulative Average Power Draw Normalized to Quantity of Ice Produced - Top Mount Unit Test #1

A similar analysis was carried out using the test data for the third ice making test, again the same level of energy stability was achieved after four compressor cycles, although this data encompassed 9 ice ejection cycles and approximately 3 hours and 18 minutes of clock time. Examination of the energy normalized to the amount of ice produced showed that the unit achieved stability much sooner than in the previous data set. Here, 1 % stability was achieved in 9 compressor cycles, or about 6 hours. Therefore this data set indicates that a test period on the order of 6 hours may be sufficient to reliably characterize the ice making energy consumption for this unit under these conditions.

This type of analysis could not be identically performed on the second data set because this data set consisted of only one compressor cycle that was not affected by a defrost sequence. The compressor on period in this data set is much longer than seen in the other data sets, since the unit was maintaining a colder set of temperatures. Figure 3.9 shows this data set plotted both as instantaneous power (pink) and cumulative average power (blue) versus time for this cabinet under the coldest setting. In total, 11 ice-ejection cycles occurred during this compressor cycle, and the time elapsed was comparable to the amount of time that it took to achieve a stable result during the other tests. It is obvious from this figure that the average running power is much more influenced by the compressor off period than the ice maker ejection heaters, therefore it is not possible to determine whether the per cycle average power is an accurate representation of steady state operation.

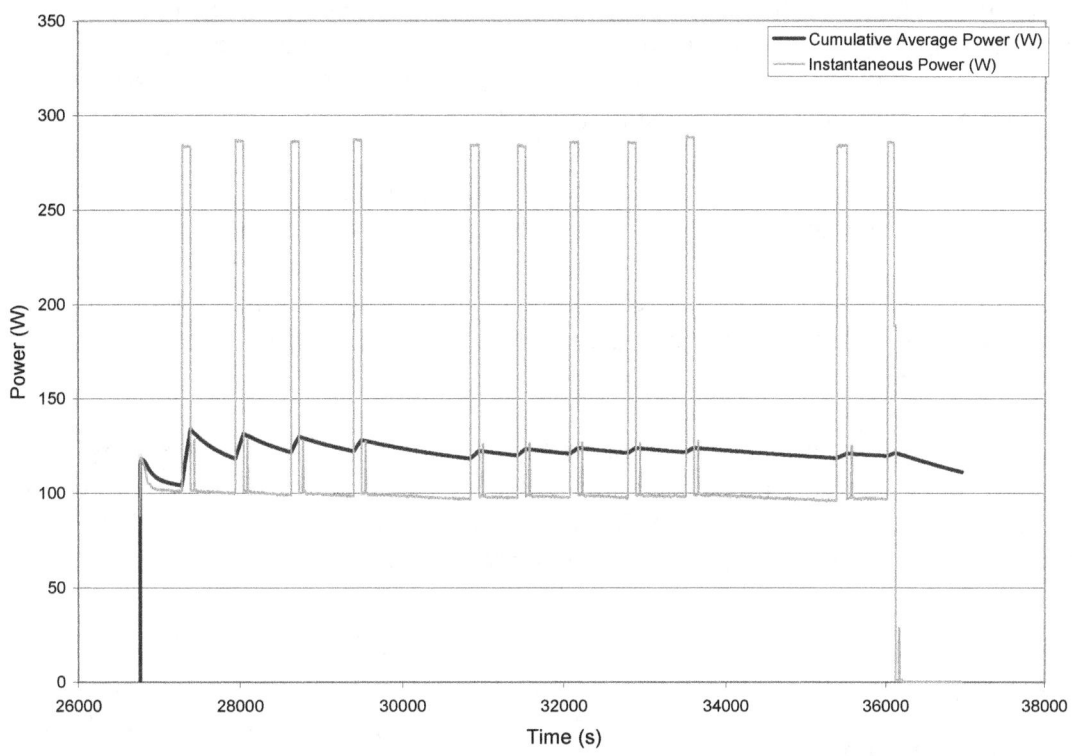

Figure 3.9 Instantaneous and Cumulative Average Power Draw
Top Mount Unit Test #2

3.5 Summary
The ice-making energy was calculated by 4 methods as shown in Table 3.1. Also shown in the table is the energy impact of producing a representative consumer usage factor of 0.816 kg of ice per day (AHAM, 2009). It should be noted that this usage factor was based on a very limited sample of survey data and will likely be revised after further studies have been completed.

The first method was to compare the difference in energy consumed with the thermostats set to their median positions. This method resulted in a large departure from the non-ice producing freezer temperature because of the control system used in this cabinet. The freezer temperature in this compartment was more than 3 °C warmer during the ice production process, which effectively biased the energy consumption results because it reduced the cooling load during the ice making test.

The second method attempted to characterize the ice-making energy by using the two-point interpolation method for two energy consumption measurements, and comparing these results to the baseline energy test results without accounting for the defrost energy. This method attempted to mitigate the effects of the elevated freezer temperature during ice production. Because of the nature of the thermostat used in this system, it was very difficult to control the temperatures during the tests, and this interpolation method resulted in a value that would represent the energy required to make ice while maintaining the appropriate freezer temperature and a refrigerator temperature that was

21

more than 3 °C colder than the target temperature. The fact that these results are interpolated to a refrigerator temperature that is so much colder than the specified temperature effectively biased the ice-making energy consumption results to show unfair penalty on the ice-making energy consumption.

The third method attempted to characterize the ice-making energy by using the three-point triangular interpolation method. The controllability of the cabinet made it impossible to actually triangulate around the specified set point; however, extrapolation of these data points should result in a reasonable estimate of the ice-making energy, depending on the linearity of the relationship between ice production rates and compartment temperatures.

The final method was a simple calculation method based on the amount of energy required to process the ice-making load, and to power the auxiliary devices required to make and store ice. This method relies heavily on an assumed COP value which is a parameter that is not readily available for a given test subject.

Table 3.1 Ice Making Energy by Various Methods – Top Mount Refrigerator Freezer				
	Method	Ice making energy		
		kWh/kg	kWh/yr	% of baseline energy
1	Median position only	0.355	106	22%
2	Median and coldest positions	0.441	132	27%
3	Triangular Interpolation	0.327	98	20%
4	Calculation	0.343	102	21%

A point for consideration is the ice maker's influence on the defrost controller. Alteration of the defrost control period will have influence the overall energy consumption. There is a wide variety of input that the current technology may use for variable automatic defrost controllers. The input used can range from measurement of ice on the evaporator, internal compartment temperatures, ambient temperatures, air humidity levels, door openings, data collected during previous operation, etc.; therefore, some defrost control algorithms may be strongly affected by the operation of ice makers and some may be completely unaffected. This particular unit was strongly influenced by the ice maker's operation, but this effect was not considered in this analysis. This effect is worthy of investigation, however, and it should be revisited at some point in the future.

After examining the energy impact of ice maker use, we examined the individual data sets to determine the minimum amount of data required to adequately characterize the ice making energy. In two of the three data sets consisting of a whole number of compressor cycles, a $\pm$ 1 % stability on ice making energy consumption was achieved in approximately 10 hours. The data set acquired at the coldest setting consisted of only one useable compressor cycle; therefore this approach could not be verified in the same manner.

4: Experimental Results for Side-by-Side Refrigerator Freezer

This unit is a 760 liter (27 cubic foot) side-by-side, energy star rated model. It has variable defrost control, electronic thermostat control, and a single-speed compressor. The unit has a factory installed ice maker, which is connected to an external water supply and produces and stores ice in a bin located on the inside of the door to the freezer compartment. This ice maker was equipped with an ON/OFF switch and is capable of producing ice at three different rates. For the purposes of these tests, all non-ice production test data were recorded with the ice maker switch set to the "ON" position and the bail arm forced to sense a full bin condition. The ice production rate was set to the highest setting.

The unit was outfitted with three thermocouples in the refrigerator compartment and five thermocouples in the freezer compartment, as shown in Figure 4.1. The compartment temperatures used for the calculations that follow are the average of the time averaged values reported from each thermocouple during the test duration.

Figure 4.1 Side-by-Side Refrigerator Freezer

The manufacturer provided the minimum and maximum time between defrosts for this unit as 7 hours and 50 hours of compressor run time, respectively. According to the method outlined in the current DOE energy test procedure, these values result in a calculated average time between defrosts of 22.4 hours of compressor run time.

4.1 Side-by-Side Refrigerator Freezer with No Ice Production

4.1.1 Mid Setting Results

This unit's electronically controlled thermostat was capable of setting the freezer temperature between -21.1 °C and -13.3 °C, and the refrigerator temperature between 0.6 °C and 7.8 °C. During the first measurement, each compartment's thermostat was set to its median temperature setting, -17.2 °C for the freezer compartment and 3.9 °C for the refrigerator compartment. Examination of the test data showed that the temperatures in the refrigerator and freezer compartment temperature changes were not in sync; i.e. the refrigerator compartment would get colder and the freezer got warmer and vice versa. This suggests that this unit supplies cold air to the compartments in an alternating pattern. From an energy test point of view, this feature showed itself as a series of secondary cycles (both power and temperature) that occurred while the compressor was operating. For the purposes of these tests, each period during which the compressor was running was considered as a single compressor "ON" cycle, regardless of the number of times it shifted the cooling load between compartments. The following results were obtained:

Steady state cyclic operation time: 14290 seconds = 03:58:10
Measured refrigerator compartment temperature: (4.2 ± 0.1) °C
Measured freezer compartment temperature: (-17.7 ± 0.1) °C
Energy expended during the test period: (401.1 ± 2.0) watt-hours

Alternatively, this can be expressed as an average steady state power of (101.1 ± 0.5) watts at the measured temperatures.

The defrost sequence for this unit consisted of (a) defrost heater switching on, (b) defrost heater switching off, (c) period of no power draw, (d) compressor on – recovery cycle, and (e) the following compressor off cycle. Measurements during this sequence yielded the following results:

Defrost time: 9355 seconds = 02:35:55
Energy expended during defrost period: (399.9 ± 2.0) watt-hours

Combining the results of these two portions of the test, assuming that the average time between defrost is 22.4 hours, we arrive at the total energy consumption of (912 ± 4.6) kWh/yr at the measured temperatures.

4.1.2 Cold Setting Results

Since the temperatures in both compartments for this test were warmer than the target temperatures, the thermostat was set to the coldest setting for each compartment, 0.6 °C for the refrigerator and -21.1 °C for the freezer, and the measurements were performed again. The power signature from this data set and the temperatures recorded in the refrigerator and freezer compartment are shown in Figures 4.2 and 4.3.

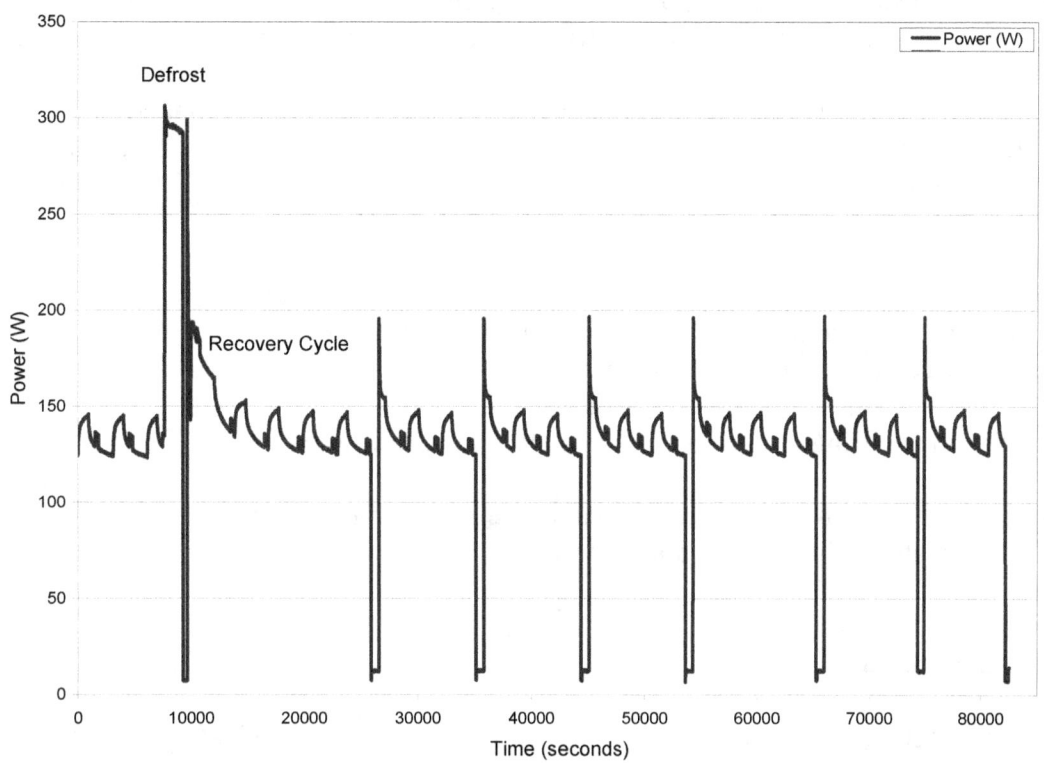

Figure 4.2 Power Signature of Side-by-Side unit - Cold/Cold Setting

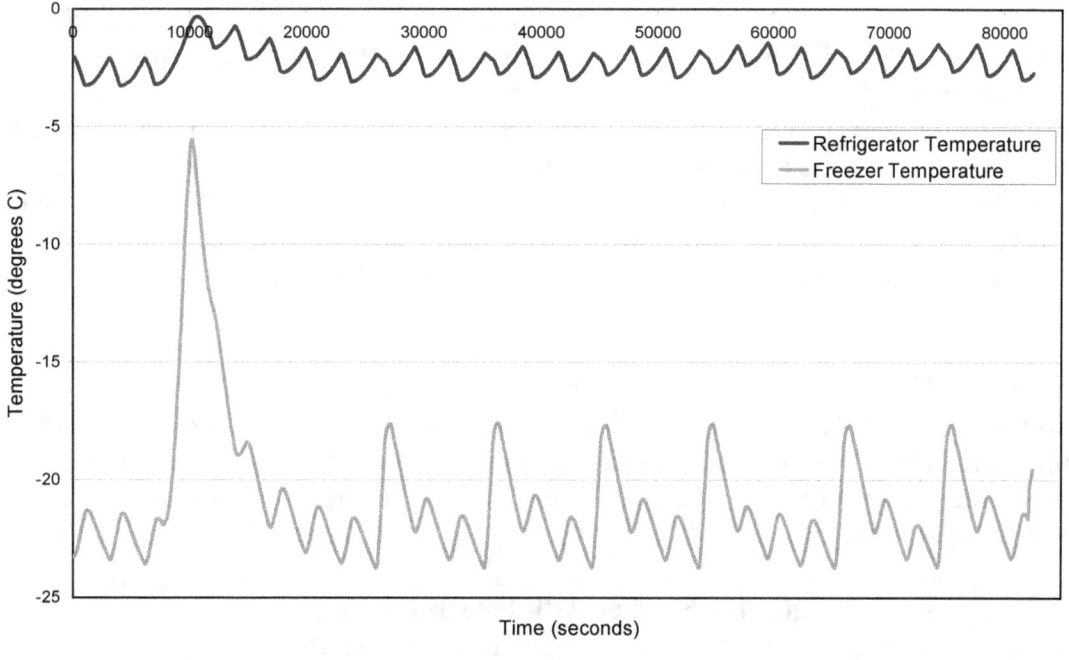

Figure 4.3 Temperature Response of Side-by-Side unit - Cold/Cold Setting

By examining the temperature response to the compressor power draw, we can see that the unit shifts its cooling load between refrigerator and freezer compartments a number of times during each compressor "ON" cycle. The cyclic operation shows that the electrical load increases and the freezer temperature increases while cooling is supplied to the refrigerator compartment, and that the electrical load decreases and the refrigerator temperature increases while it is not.

The following results were obtained:

Steady state cyclic operation time: 20715 seconds = 05:45:15
Measured refrigerator compartment temperature: (-2.0 ± 0.1) °C
Measured freezer compartment temperature: (-21.5 ± 0.1) °C
Energy expended during the test period: (729.9 ± 3.6) watt-hours

Alternatively, this can be expressed as an average steady state power of (126.8 ± 0.6) watts at the measured temperatures.

Defrost time: 18882 seconds = 05:14:42
Energy expended during defrost period: (776.8 ± 3.9) watt-hours

These values resulted in the total energy consumption of (1133 ± 6) kWh/yr at the measured temperatures.

4.1.3 Combined Test Results for Side-by-Side Refrigerator with No Ice Production

The results of the measurements at the first thermostat setting were combined with the results at the second thermostat setting and interpolated to the target temperatures. When interpolating to the refrigerator target temperature of 3.9 °C, the calculated energy consumption was (923 ± 3) kWh/yr; when interpolating to the freezer target temperature of -17.8 °C, the energy consumption was (918 ± 6) kWh/yr. The maximum of these two values is declared as the baseline energy consumption: (923 ± 3) kWh/yr. It is important to note that the interpolation results using the refrigerator temperature target and that using the freezer temperature target are very close to each other, therefore using the triangular interpolation method will not result in improved accuracy.

It is also useful for later calculations to note the energy consumption obtained through interpolation if the defrost energy was not considered. In this case, the calculations were performed with the steady state operational average power of (101.1 ± 0.5) W (mid/mid setting) and (126.8 ± 0.6) W (cold/cold setting) rather than the total values. Here, negating the energy from periodic defrosting shows the energy consumption for this unit to be (897 ± 3) kWh/yr.

4.2 Side-by-Side Refrigerator Freezer with Ice Production

With the baseline energy consumption characterized, we next performed a series of measurements to examine the energy consumption attributed to the production of ice by the automatic ice maker.

4.2.1 Mid Setting Results

The thermostat was set to the median position for each compartment and the ice maker was rendered active. Water was supplied to the unit at a temperature of 32.2 °C.

Figures 4.4 and 4.5 show the power and temperatures measured during the test. These figures demonstrate the complexity of the operation of this unit. The first point to note is that the compressor cyclic pattern changes periodically, this is because of the nature of the electronic thermostatic controller used in this system. This controller monitors the temperatures in both compartments and supplies cooling to each one based on its needs. Satisfying both of these compartments needs simultaneously results in the operational mode switching as shown in the figures.

The power signature of this unit also shows the operational energy consumed by the ice making components. The ice ejection heater operates at approximately 130 watts for a period of 150 seconds (19.5 kJ). After the ice is freed from the mold, a plastic rod is swept through each mold and pushes the frozen cubes to the storage bin. Then the solenoid valve opens to deliver water to the molds; this process consumes approximately 32 watts for 4 seconds (128 J).

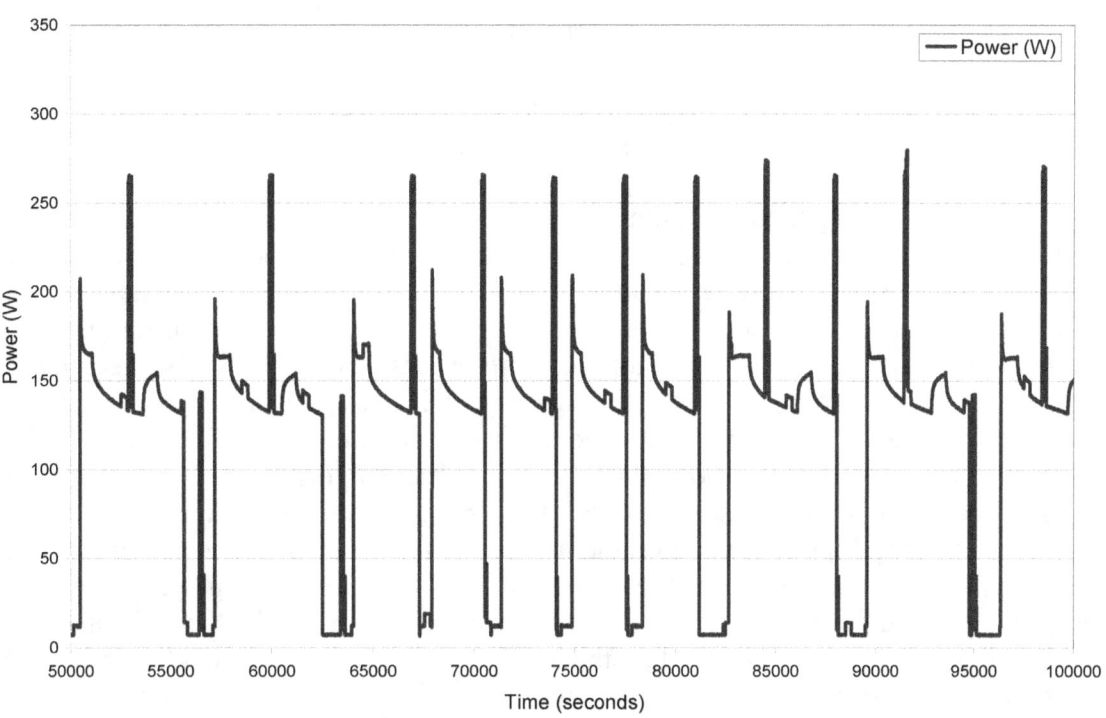

Figure 4.4 Refrigerator Power at Mid/Mid Setting with Ice Production
for Side-by-Side Refrigerator Freezer

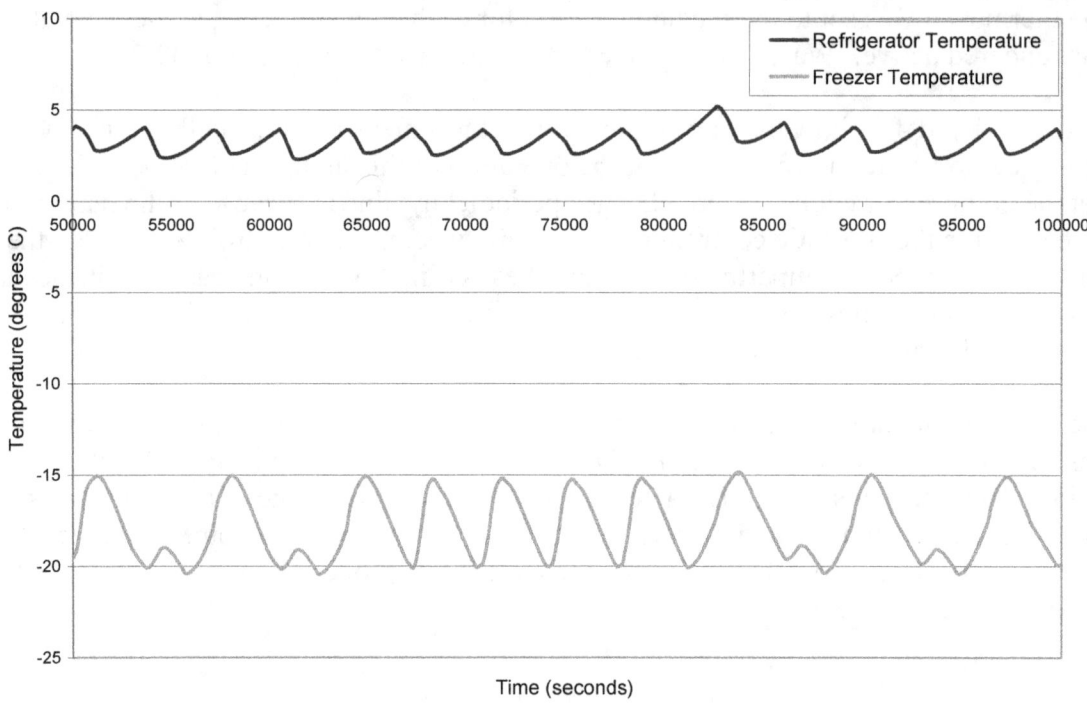

Figure 4.5 Temperature Response to Mid/Mid Setting with Ice Production
for Side-by-Side Refrigerator Freezer

It is also very interesting to note that the programming algorithm for the ice maker forces it to cycle at regular intervals. This is not a very good approach since water freezes into ice at different rates depending on the amount of cold air supplied to the ice maker, which is influenced by the compressor on/off cycles. The problem with this algorithm is that it often undergoes the ice ejection and mold refilling process without actually checking whether the water has frozen into ice. In other words, if the water in the mold has not been frozen, the ejection heater adds heat to that water and the plastic ejection rod will find no cube to push out of the mold. Overall, this control algorithm greatly reduced the amount of ice actually produced by this unit; this unit generally yields approximately 1/3 of the amount of ice expected based on the number of ice-making cycles and number of cube molds. It is important to note the possibility that these ice-making cycles may be appropriately timed for domestic use (25 °C environment), but not for the current 32.2 °C ambient test conditions as evidenced by the low ice yield. It is also important to note that if the timing of these cycles is not appropriate for normal use, the user would not realize the defect without examining the power signature because the unit would still produce whole ice cubes.

The data was recorded for the maximum amount of time that could be logged while avoiding a defrost cycle. The data was also reduced to allow a whole number of compressor cycles; the ice production cycles were again treated as secondary.

Steady state cyclic operation time: 126030 seconds = 35:00:30

28

Measured refrigerator compartment temperature: $(3.2 \pm 0.1)\ ^{\circ}C$
Measured freezer compartment temperature: $(-18.1 \pm 0.1)\ ^{\circ}C$
Mass of ice produced: (0.85 ± 0.005) kg
>   calculated based on the assumption of constant mass of ice produced per cycle, 1.51 kg of ice produced in 64 ice making cycles, and counting 36 ice making cycles occurring during the noted test period

Ice production rate: (24.27 ± 0.14) grams per hour
Energy expended during the test period: (4.18 ± 0.02) kWh
Average Power drawn during test period: (119.3 ± 0.6) W

Single Data Point Method for Measuring Ice-Maker Energy Consumption
The first proposed approach to quantifying the ice-making energy is to compare the steady state energy consumption with both thermostats set to their median setting for both ice producing and ice maker inoperative conditions. Since this particular unit had an electronic thermostat, the compartment temperatures measured during the tests at the median temperature settings were similar regardless of whether the ice maker was producing ice or not; therefore, it is thought that this method should provide a fairly good comparison. In this case, the differential power consumption was:

$$(119.3 \pm 0.6)\ W - (101.1 \pm 0.5)\ W = (18.2 \pm 0.8)\ W$$

The energy used to produce ice in this configuration can then be calculated from the mass of ice, the differential power and the test period time.

$$\frac{(0.0182\ kW)(35\ h)}{(0.85\ kg)} = (0.749 \pm 0.03)\ kWh/kg$$

4.2.2 Warm Setting Results
The next approach was to attempt to use an interpolated energy (and ice production rate) using a second set of measurements. Since the internal temperatures of both compartments were colder than the target temperatures, the thermostats of each compartment were set to the warmest setting for the second measurement. Again, the data was logged for the longest attainable period during which ice was actively produced without the influence of the defrost cycle. The following results were obtained:

Steady state cyclic operation time: 232849 seconds = 64:40:49
Measured refrigerator compartment temperature: $(8.1 \pm 0.1)\ ^{\circ}C$
Measured freezer compartment temperature: $(-13.3 \pm 0.1)\ ^{\circ}C$
Mass of ice produced: (1.23 ± 0.005) kg
>   calculated based on 1.37 kg of ice produced in 68 ice making cycles and counting 61 ice making cycles occurring during the noted test period

Ice production rate: (19.02 ± 0.08) grams per hour
Energy expended during the test period: (6.22 ± 0.03) kWh
Average Power drawn during test period: (96.11 ± 0.48) W

Two Data Point Method for Measuring Ice-Maker Energy Consumption

Now, the energy consumption at the specified target temperatures is calculated based on the results of the mid/mid and warm/warm setting tests. For this calculation, the results show that the interpolated values for a freezer temperature of -17.8 °C are an energy consumption of (1032 ± 4) kWh/yr and an interpolated ice production rate of (210 ± 1) kg/yr. The interpolation results using the refrigerator compartment temperatures are again very close to the interpolation results using the freezer temperatures. Comparing these values to those calculated from the tests without ice production (not considering defrost periods) results in the following values:

Energy consumption difference:
(1032 ± 4) kWh/yr - (897 ± 3) kWh/yr = (135 ± 5) kWh/yr
Ice production rate: (210 ± 1) kg/yr
Ice production energy: (0.643 ± 0.02) kWh/kg

4.3 Calculation Based Method

The next method of ice maker energy to be examined is a simplified calculation approach. Considering that this particular unit yields far less ice per ice making cycle than expected, we will examine this approach by using both the actual ice production rate and a theoretical ice production rate based on the assumption that each ice cube mold produces one cube during every ice-making cycle.

Each data set taken with the ice maker operative was started with the ice bin empty and logged until the bin reached its shut off limit. The first data set showed 1.51 kg of ice produced through 64 cycles, the second showed 1.37 kg in 68 cycles. This large difference in production rates is due to the fact that this unit harvests more ice when the freezer cabinet is colder, yet seems to use relatively similar amounts of power because of the regularity of the ice production cycle as described earlier. For the purposes of these calculations, summing the results of these two tests yields a production rate of 21.8 grams of ice per cycle (2.88 kg in 132 cycles).

As mentioned earlier, the solenoid valve consumes 128 Joules per cycle, or 5.87 kJ/kg of ice produced. The ice ejection heater consumes 19.5 kJ, or 894.5 kJ/kg, which is then dissipated into the freezer compartment. Therefore, the total energy consumed by these components is 900.4 kJ/kg of ice produced. Considering the heat load for freezing water into ice and the removal of the dissipated heat from the ejection heaters, the total demand of additional cooling to the refrigeration system is 1405.4 kJ/kg. Assuming that the refrigerating system operates with a COP of 1.8, this translates to 781 kJ/kg. Finally, adding in the energy consumed by the solenoid, ejection rod, and ejection heater brings the total load to 1681 kJ/kg; or in alternative units 0.467 kWh/kg.

The theoretical ice production rate for this unit is approximately 60 grams of ice per cycle, based on the ice mold producing batches of 6 cubes of 10 grams each per ice making cycle. This is a substantially larger ice production rate than the actual rate, which highlights the inefficiency of this unit under the current test conditions and emphasizes the rationale for avoiding a purely calculation based approach. Using this theoretical ice production rate, the solenoid valve and cube removal mechanism consume 2.13 kJ/kg of ice, and the ejection heater consumes 325 kJ/kg of ice, a total of 327.1 kJ/kg of ice. The combined heat load of cooling and freezing of water plus the heat addition from the ejection heater totals 830 kJ/kg, which is removed using the refrigeration system with the assumed COP of 1.8, yielding additional energy consumption of 461 kJ/kg. Adding back in the energy from the solenoid, ejection rod, and ejection heater brings the total load to 788 kJ/kg; or in alternative units 0.218 kWh/kg. This value is clearly a gross misrepresentation of the actual energy used to produce ice under these conditions; however it is noted to demonstrate the dangers of a poorly defined calculation based method.

4.4 Test Length Determination

As with the previous test subject, we examined the power consumed by the unit while making ice in a little more detail in order to determine how much data needs to be recorded to properly calculate the energy consumption. Figure 4.6 shows the average power consumed during each full compressor cycle while the unit was operating at the mid/mid setting and producing ice.

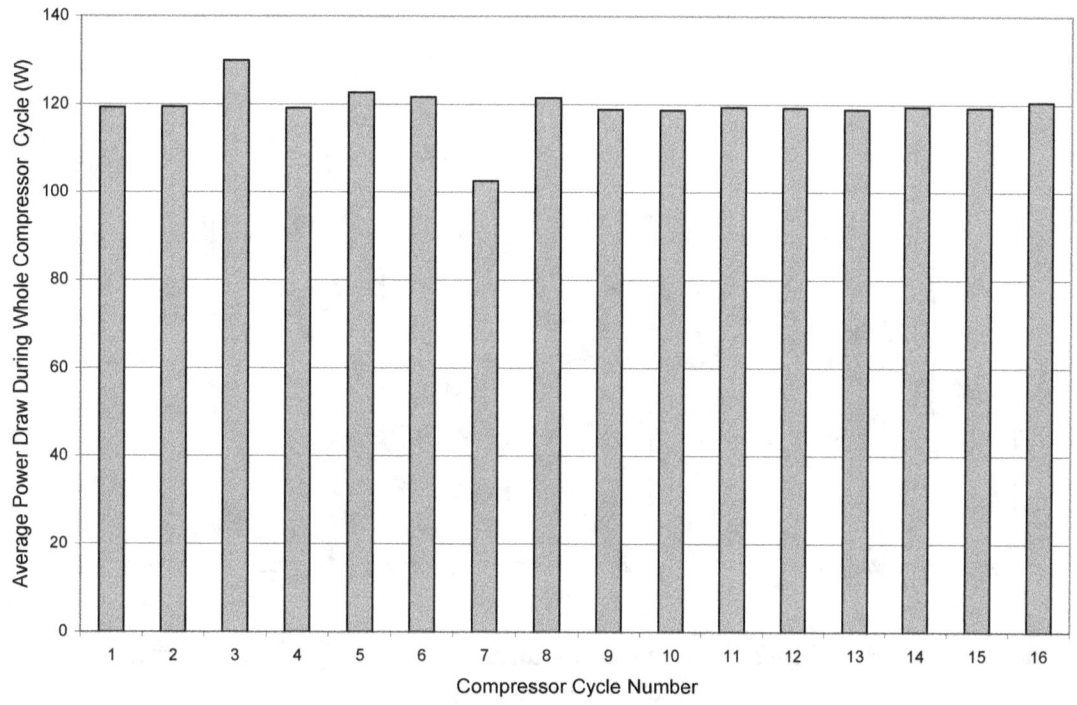

Figure 4.6 Average Power Draw Over Full Compressor Cycle – Side by Side Unit Test #1

In this set of data, the ice maker cycled at fixed time intervals but the compressor cycle was somewhat irregular due to the frequent shifting of the cooling load between the refrigerator and freezer compartment. Examination of the data, however, shows that the cumulative average power becomes stable to within 1 % of the final value after the 7th compressor cycle. When examining the amount of energy consumed normalized to the amount of ice produced, however, this stability metric is much worse. In this case, the

31

1% stability criterion was achieved after 22 compressor cycles, encompassing approximately 31 hours of test time. The unit achieved 2 % stability in 21 compressor cycles (30 hours) and 4 % stability in 17 compressor cycles (26 hours)

Figure 4.7 shows the data plotted as both instantaneous power (pink) and cumulative average power draw (blue) versus time for this cabinet under the warm/ warm thermostat setting. We searched for the point where the cumulative running average of energy consumption changes by less than 1 % between successive compressor cycles. The results showed that it is reasonable to take the first 7 compressor cycles, which encompasses 9 ice ejection cycles and nearly 9 hours of data, as a test period that would produce a good energy consumption result. Again, this lengthy period is due to the fact that this unit operated some energy consuming features intermittently, which influenced the cumulative average power draw stability metric. Had these features been inoperative, this unit would have satisfied the 1 % criteria much sooner.

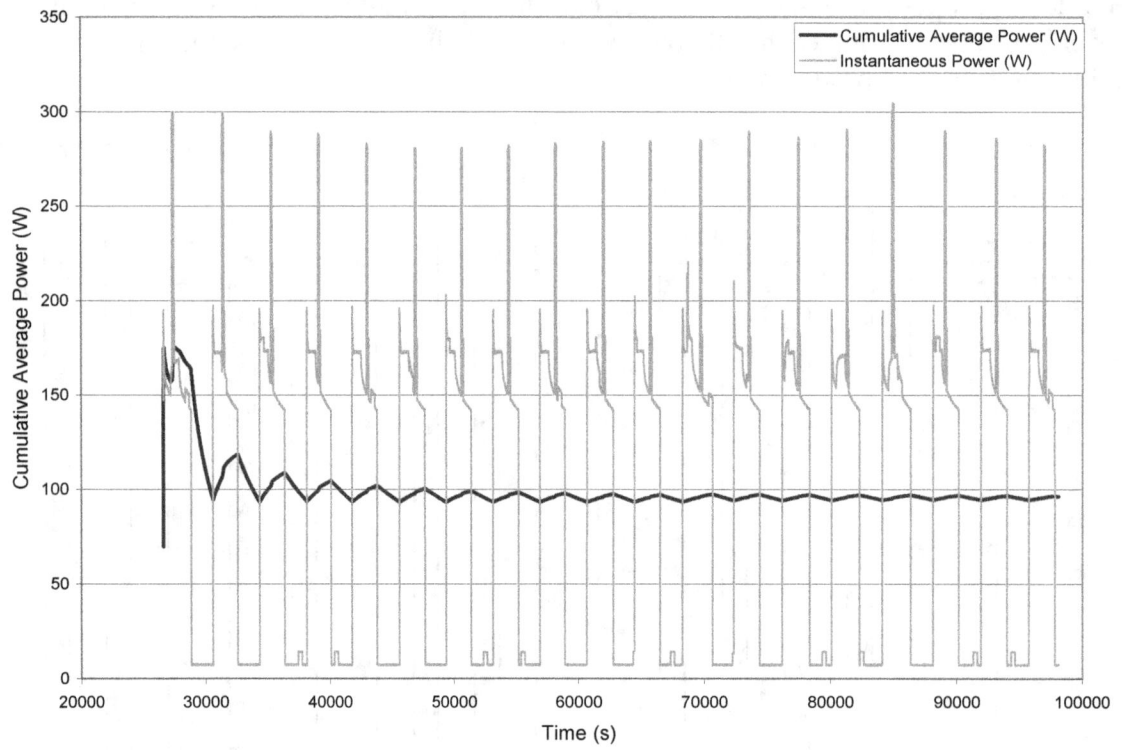

Figure 4.7 Instantaneous and Cumulative Average Power Draw
Side-by-Side Unit Test #1

An analysis of the ice making energy to quantity of ice produced was performed using the second data set. This data set showed that the unit achieved 1% stability within 12 compressor cycles under these conditions. This period encompassed 12 ice making cycles and nearly 13 hours of clock time. This is better than the data for this unit at the mid/mid thermostat settings because of the regular compressor cycle and the synchronous ice maker cycles.

4.5 Summary

Table 4.1 shows the ice making energy was calculated by 3 methods. Also shown in Table 4.1 is the annual energy impact of producing 0.816 kg of ice per day.

Table 4.1 Ice Making Energy by Various Methods – Side by Side Refrigerator Freezer				
Method		Ice making energy		
		kWh/kg	kWh/yr	% of baseline energy
1	Median position only	0.749	223	24%
2	Two point interpolation	0.643	192	21%
3	Calculation	0.467	139	15%
4	Purely theoretical calculation	0.218	62.5	6.8%

5: Experimental Results for French Door Refrigerator Freezer #1

This unit is an 800 liter (28 cubic foot) French door, energy star rated model with through the door ice and water service. It has variable defrost control, electronic thermostat control, and a variable-speed compressor. The unit has a factory installed ice maker which is located in the upper portion of the refrigerator compartment. The ice maker is connected to an external water supply and produces and stores ice in a bin also located in the refrigerator compartment. The ice maker for this unit has a dedicated evaporator and produces ice without displacing cold air from the freezer compartment. The unit was outfitted with three thermocouples in the refrigerator compartment, as shown below, and five thermocouples in the freezer compartment. The compartment temperatures used for the calculations that follow are the average of the time averaged values reported from each thermocouple during the test duration.

Figure 5.1 French Door Unit #1

The minimum and maximum time between defrost for this unit are 6 hours and 96 hours of compressor run time, respectively. According to the procedure outlined in the current

DOE energy test procedure, these values result in a calculated average time between defrost of 24 hours of compressor run time.

This unit also has an automatically controlled anti-sweat heater, which responds to the ambient humidity condition. The humidity levels were maintained below 8 °C dew point for all tests performed in this study in order to ensure that this feature would not consume electricity outside the scope of this set of experiments.

5.1 French Door Refrigerator Freezer #1 with No Ice Production
5.1.1 Mid Setting Results
During the first test, the energy consumption was measured at with the thermostat set to the median setting for each compartment. The fact that this unit had a variable-speed compressor facilitated the characterization of energy consumption because the compressor did not undergo periodic cycles; rather it maintained part load continuous operation to stabilize the temperatures in the cabinets. The following results were obtained:

Steady state operation time: 40145 seconds = 11:09:05
Measured refrigerator compartment temperature: (6.9 ± 0.1) °C
Measured freezer compartment temperature: (-17.9 ± 0.1) °C
Energy expended during the test period: (629.2 ± 3.1) watt-hours

Alternatively, this can be expressed as an average steady state power of (56.42 ± 0.28) watts at the measured temperatures. In alternative units, (494.3 ± 2.5) kWh/yr.

The defrost sequence for this unit consisted of (a) defrost heater switching on, (b) defrost heater switching off, (c) period of no power draw, (d) a recovery cycle – high compressor power, and (e) return to steady state – part load compressor power. The sequence is shown below in Figure 5.2.

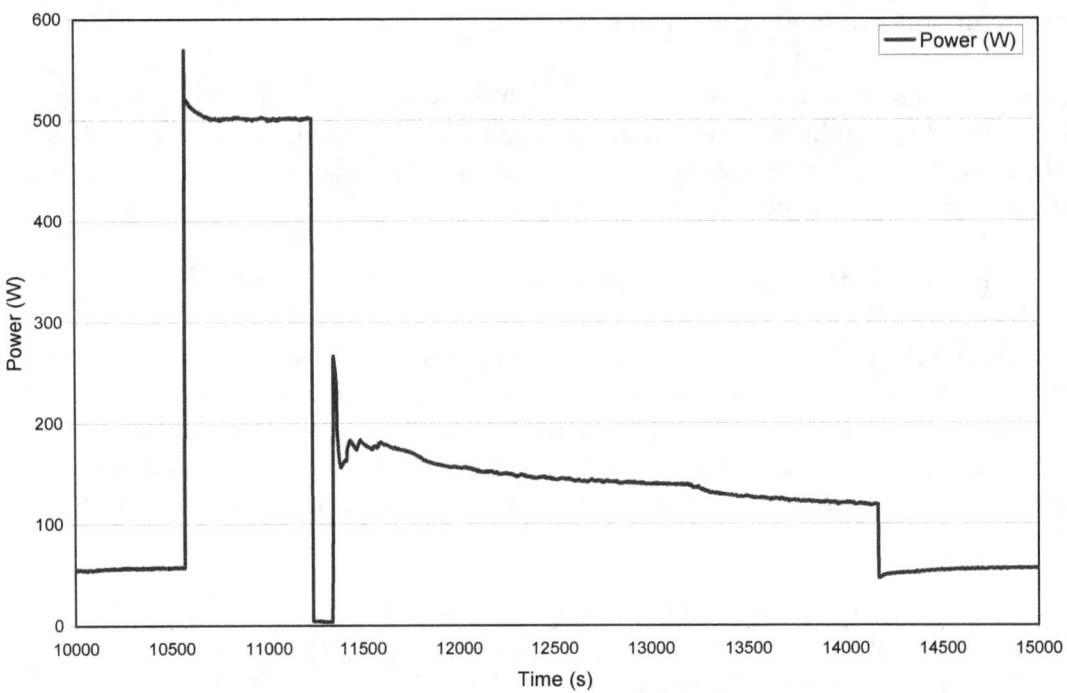

Figure 5.2 Refrigerator Power during Defrost – French Door Unit #1

Measurements during this sequence yielded the following results:

Defrost time: 3598 seconds = 00:59:58
Energy expended during defrost period: (205.3 ± 1.0) watt-hours

Combining the results of these two portions of the test, assuming that the average time between defrost is 24 hours, we arrive at the total energy consumption of (521.5 ± 2.6) kWh/yr at the measured temperatures. It should be noted that the calculations used to arrive at this number are based on an assumption that the compressor operates 50% of the time; which is not valid for a variable speed compressor which operates 100% of the time under part load conditions.

5.1.2 Cold Setting Results
The second set of tests were conducted with the thermostat for each compartment set to the coldest setting. The following results were obtained:

Steady state operation time: 89901 seconds = 24:58:21
Measured refrigerator compartment temperature: (4.4 ± 0.1) °C
Measured freezer compartment temperature: (-22.7 ± 0.1) °C
Energy expended during the test period: (1756.4 ± 8.8) watt-hours

Alternatively, this can be expressed as an average steady state power of (70.33 ± 0.35) watts at the measured temperatures, or (616.1 ± 3.1) kWh/yr.

Measurements during this sequence yielded the following results:

Defrost time: 5426 seconds = 01:30:26
Energy expended during defrost period: (266.7 ± 1.3) watt-hours

Incorporating the defrost energy results in an overall energy consumption value of (645.4 ± 3.2) kWh/yr at these temperatures.

5.1.3 Mixed Setting Results
Since the thermostat for this unit was programmed to optimize performance while simultaneously maintaining temperatures of 7.2 °C and -15.0 °C in the refrigerator and freezer, respectively, a third data set was taken with the refrigerator set to the coldest setting and the freezer set to the warmest setting. The third data set resulted in the following values:

Steady state operation time: 39733 seconds = 11:02:13
Measured refrigerator compartment temperature: (2.4 ± 0.1) °C
Measured freezer compartment temperature: (-15.7 ± 0.1) °C
Energy expended during the test period: (673.5 ± 3.4) watt-hours

Alternatively, this can be expressed as an average steady state power of (61.0 ± 0.3) watts at the measured temperatures, or (534.5 ± 2.7) kWh/yr. Measurements during this sequence yielded the following results:

Defrost time: 4304 seconds = 01:11:44
Energy expended during defrost period: (250.9 ± 1.3) watt-hours

Incorporating the defrost energy results in an overall energy consumption value of (567.0 ± 2.8) kWh/yr at these temperatures.

5.1.4 Combined Test Results for French Door Refrigerator Freezer #1 with No Ice Production
The steady operation values from these three data sets were used to calculate the energy consumption at -17.8 °C and 3.9 °C using the three point interpolation method; this resulted in an energy consumption value of (544 ± 5) kWh/yr. The total energy consumption, including the defrost energy, was (574 ± 5) kWh/yr.

5.2 French Door Refrigerator Freezer #1 with Ice Production
With the baseline energy consumption characterized, we next performed a series of measurements to examine the energy consumption attributed to the production of ice by the automatic ice maker.

5.2.1 Mid Setting Results
First, the thermostat was set to the median position for each compartment and the ice maker was rendered operative. Water was supplied to the unit at a temperature of 32.2 °C. Figure 5.3 shows the power signature of this unit while making ice under these conditions.

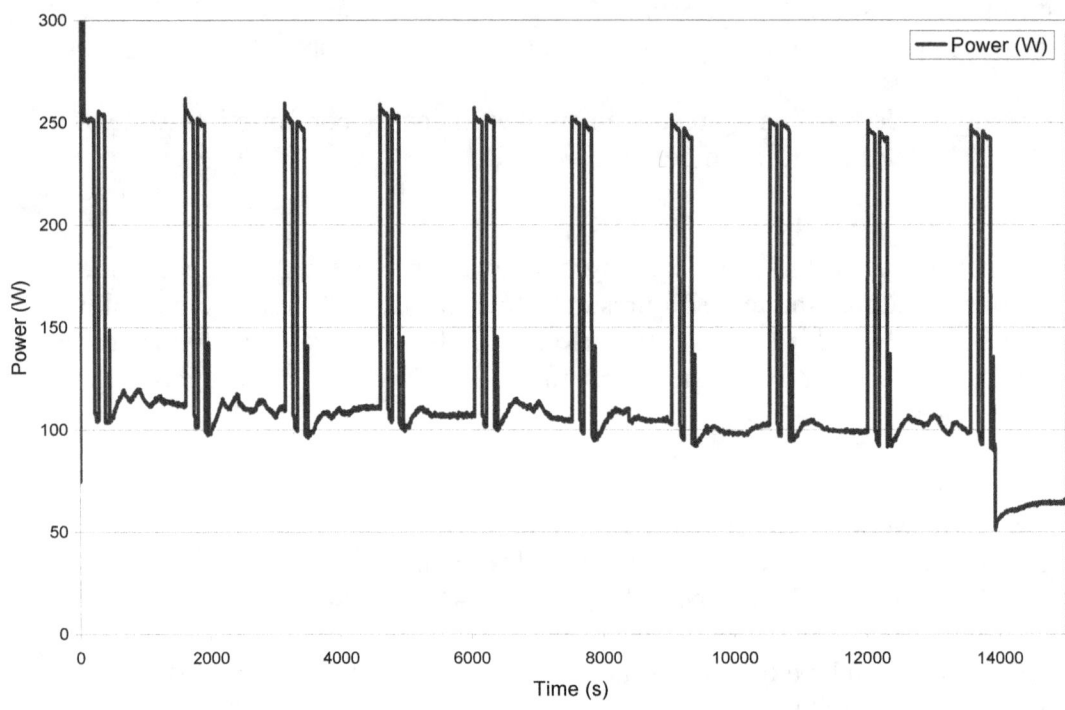

Figure 5.3 Refrigerator Power During Ice Production at Mid/Mid Setting for
French Door Unit #1

Starting from an empty ice bin condition, the ice maker operated for less than 14 000
seconds before ceasing ice production by sensing a full bin. During this ice production
period, the freezer temperature deviated from its measured temperature prior to the ice
making period, shown in Figure 5.4. It is noted that each ice making cycle appears to
include two successive ejection heater operation periods followed by the solenoid
operation. Each time the heater is powered on, it consumes 150 watts for approximately
120 seconds; the solenoid consumes 40 watts for 4 seconds. It is also noted that the
freezer compartment became slightly colder for a short period, which was followed by a
period of increasing freezer temperature. This is documented merely to illustrate that the
ice making period is a transient period during which time the unit tries to adjust operation
to produce ice while maintaining the prescribed conditions. For this reason, it was
generally not possible to acquire a data set that encompasses real steady state operation.

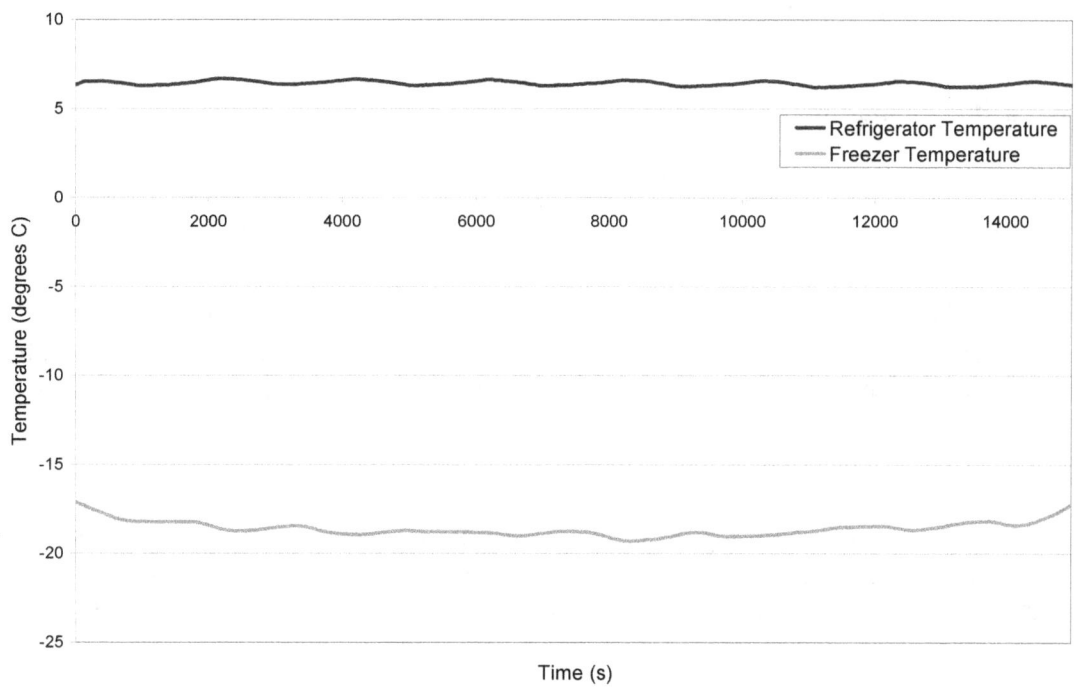

Figure 5.4 Temperature Response During Ice Production at Mid/Mid Setting for
French Door Unit #1

Since this unit had a variable-speed compressor, it was not necessary to consider
compressor cycles, and the ice making energy was calculated by the considering all of the
ice making cycles. The test period is then described as starting from the ice maker being
cued to produce ice until the moment that the ice maker halts production based on it
sensing a full bin condition. This is clearly visible on the figure above showing the
power signature.

Operation time: 13920 seconds = 03:52:00
Measured refrigerator compartment temperature: (6.2 ± 0.1) °C
Measured freezer compartment temperature: (-18.8 ± 0.1) °C
Mass of ice produced: (0.69 ± 0.005) kg
Ice production rate: (178.5 ± 1.3) grams per hour
Energy expended during the test period: (503.0 ± 2.5) watt-hours

Alternatively, this can be expressed as an average steady state power of
(130.1 ± 0.7) watts at the measured temperatures; or (1139.5 ± 5.7) kWh/yr.

Single Data Point Method for Measuring Ice Maker Energy Consumption
The first proposed approach to quantifying the ice making energy is to compare the
steady state energy consumption with both thermostats set to their median setting for both
ice producing and ice maker inoperative conditions. In this case, the differential power
consumption was:

$$(130.1 \pm 0.7) \text{ W} - (56.42 \pm 0.28) \text{ W} = (73.7 \pm 0.8) \text{ W}$$

The energy used to produce ice in this configuration can then be calculated from the mass of ice, the differential power and the test period time.

$$\frac{(0.0737 \text{ kW})(3.867 \text{ h})}{(0.69 \text{ kg})} = 0.413 \text{ kWh/kg} \ (\pm 0.005 \text{ kWh/kg})$$

In this case, the measured temperature in both compartments was colder than the measured temperature during the first test (same setting but no ice production), therefore the results of this ice-making test also include the additional load of maintaining slightly colder temperatures.

5.2.2 Cold Setting Results

The next approach was to attempt to use an interpolated energy (and ice production rate) using this and two other sets of measurements. We set the thermostats of each compartment to the coldest setting for the second ice making data set. The figure below shows the power signature recorded during this measurement.

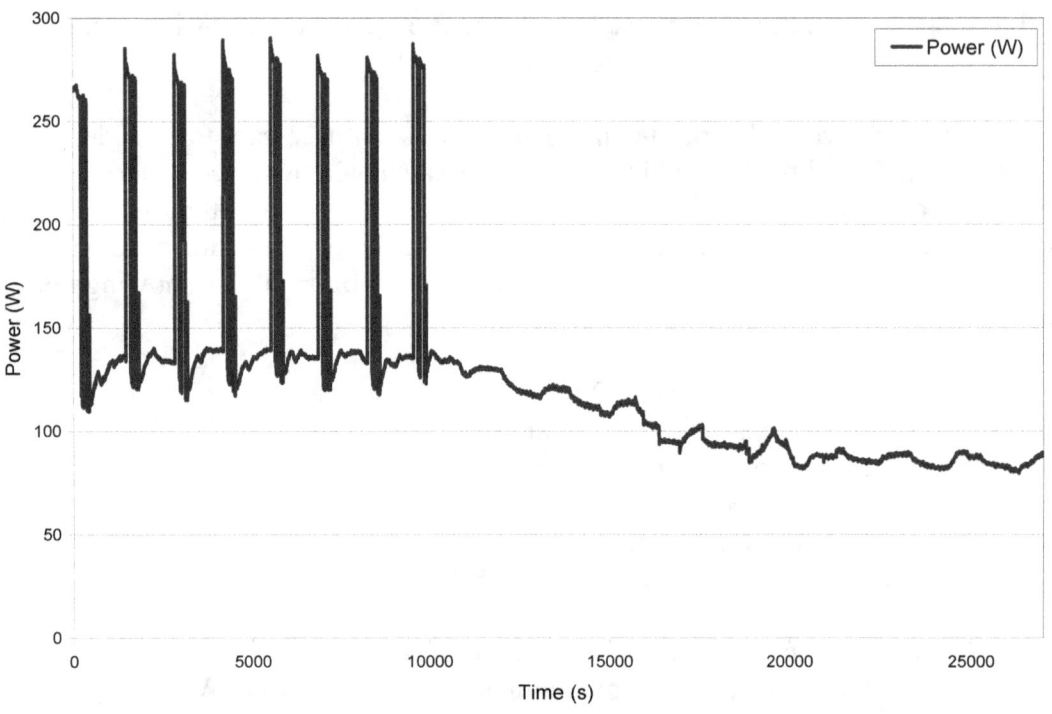

Figure 5.5 Power Response During Ice Production at Cold/Cold Setting for French Door Unit #1

Under these test conditions, the unit does not abruptly shift back to the steady operation after it ceased ice production due to the full bin condition as was the case with the previous measurement. Rather it gradually reduced the compressor power until reaching the steady operation condition. The figure shows that ice production ceases at approximately 10 000 seconds, but the unit does not recover from the ice production until approximately 20 000 seconds. If we take the test period to be that of the ice production period only, we obtain the following values:

Operation time: 9906 seconds = 02:45:06
Measured refrigerator compartment temperature: (3.5 ± 0.1) °C
Measured freezer compartment temperature: (-22.3 ± 0.1) °C
Mass of ice produced: (0.55 ± 0.005) kg
Ice production rate: (199.9 ± 1.8) grams per hour
Energy expended during the test period: (436.0 ± 2.2) watt-hours

Alternatively, this can be expressed as an average steady state power of (158.5 ± 0.8) watts at the measured temperatures; or (1388.1 ± 6.9) kWh/yr.

If the recovery period is included as part of the analysis, the following values are obtained:

Operation time: 20000 seconds = 05:33:20
Measured refrigerator compartment temperature: (3.4 ± 0.1) °C
Measured freezer compartment temperature: (-23.0 ± 0.1) °C
Mass of ice produced: (0.55 ± 0.005) kg
Ice production rate: (99.0 ± 0.9) grams per hour
Energy expended during the test period: (747.0 ± 3.7) watt-hours

Alternatively, this can be expressed as an average steady state power of (134.5 ± 0.7) watts at the measured temperatures; or (1177.8 ± 5.9) kWh/yr.

Inclusion of the recovery period resulted in a colder freezer temperature and lower energy consumption, but also lower ice production rate.

5.2.3 Mixed Setting Results
The final measurement for this unit was performed with the freezer set to the warmest setting and the refrigerator set to the coldest setting. This data set, like that from the mid/mid setting with ice production, showed an abrupt reduction in compressor speed at the time that it ceased ice production. The following data were obtained from these measurements:

Operation time: 11039 seconds = 03:03:59
Measured refrigerator compartment temperature: (2.9 ± 0.1) °C
Measured freezer compartment temperature: (-17.1 ± 0.1) °C
Mass of ice produced: (0.56 ± 0.005) kg
Ice production rate: (182.6 ± 1.6) grams per hour
Energy expended during the test period: (395.2 ± 2.0) watt-hours

Alternatively, this can be expressed as an average steady state power of (128.9 ± 0.6) watts at the measured temperatures; or (1128.9 ± 5.6) kWh/yr.

Three Data Point Method for Measuring Ice Maker Energy Consumption

By applying the three point interpolation method for both the energy consumption and the ice production rate, the following values were obtained for conditions representative of 3.9 °C in the refrigerator and -17.8 °C in the freezer. The following data shows the overall results of the calculations using both possible ice production periods from the cold/cold setting data (i.e., with and without the recovery period).

Ice production period only
Energy consumption at -17.8/3.9 = (1142 ± 7) kWh/yr = (130.4 ± 0.8) watts
Ice production rate at -17.8/3.9 = (182 ± 1) grams per hour

Ice production and recovery periods
Energy consumption at -17.8/3.9 = (1134 ± 4) kWh/yr = (129.5 ± 0.5) watts
Ice production rate at -17.8/3.9 = (179 ± 2) grams per hour

The results differ by 0.7 % on energy and 1.7 % on ice production rate. These calculations show that the result is largely unaffected by the inclusion of the recovery period for this unit; however, this may not hold true for a unit with a more aggressive recovery schedule.

The following results were obtained by comparing the triangularly interpolated results from the tests with and without ice production.

$$(130.4 \pm 0.8) \ W - (62.1 \pm 1.7) \ W = (68.3 \pm 1.9) \ W$$

The energy used to produce ice in this configuration can then be calculated from the mass of ice, the differential power and the test period time.

$$\frac{(0.0683 \ \text{kW})}{\left(0.182 \ \frac{\text{kg}}{\text{h}}\right)} = (0.375 \pm 0.011) \ \text{kWh/kg}$$

Next, it is also interesting to calculate the ice making energy by triangular interpolation of the non-ice making data to the temperatures realized during the ice making tests. First, the ice making test data at the mid/mid setting resulted in temperatures of 6.2 °C in the refrigerator and -18.8 °C in the freezer. Triangularly interpolating the non-ice making data points to these temperatures, we calculate a value of (59.48 ± 0.35) W, which is (70.6 ± 0.8) W less than the average power seen during the mid/mid test with ice production. Dividing this differential power by the ice production rate, the ice making energy at T_{fridge}=6.2 °C and T_{freeze}=-18.8 °C is (0.396 ± 0.005) kWh/kg. Similarly, the ice making energy was calculated at the coldest and mixed settings as well. The coldest setting, at T_{fridge}=3.5 °C and T_{freeze}=-22.3 °C is (0.436 ± 0.020) kWh/kg; the mixed setting, at T_{fridge}=2.9 °C and T_{freeze}=-17.1 °C is (0.363 ± 0.005) kWh/kg. These results illustrate the dependence of the ice-making energy on the compartment temperatures. It is for this reason that a fixed set of temperature conditions should be used in order to form a repeatable basis for an ice making energy test.

5.3 Calculation Based Method

Lastly, the theoretical ice making energy was calculated using the simplified calculation approach. Each data set taken with the ice maker operative was started with the ice bin empty and logged until the bin reached its shut off limit. The first data set showed 0.69 kg of ice produced through 10 cycles, the second showed 0.55 kg in 8 cycles, and the third showed 0.56 kg in 8 cycles. On average, this unit produces 69.2 grams of ice during each cycle.

The solenoid valve draws 40 watts for a period of 4 seconds to deliver water to the molds, which results in 160 Joules during each ice making cycle, or 2.31 kJ/kg of ice produced. The ice ejection heater draws 150 watt for two 120 second periods to free the ice from the molds which results in 36,000 Joules per ice making cycle, or 520.2 kJ/kg of ice produced. The total energy consumed by these components is therefore 522.5 kJ/kg of ice produced. Also, the energy consumed by the ice ejection heater is ultimately dissipated as heat into the ice making compartment; therefore it must be considered an additional heat load to be processed by the refrigeration system.

The energy required to cool 32.2 °C water to freezing accounts for 505 kJ/kg. The refrigeration system must therefore meet the additional demand to remove 1025 kJ of heat per kg of ice produced. Using the assumed COP of 1.8, the additional compressor duty to remove this heat will total 569.5 kJ/kg. Adding in the energy consumed by the solenoid and ejection heater brings the total load to 1092.0 kJ/kg; or in alternative units 0.303 kWh/kg.

5.4 Summary

The ice-making energy was calculated by 3 methods as shown in Table 5.1. Also shown in the table is the energy impact of producing 0.816 kg of ice per day.

The first method was to compare the difference in energy consumed with the thermostats set to their median positions. This method resulted in a temporary, relatively small departure from the non-ice producing freezer temperature because of the control system used in this cabinet. The second method attempted to characterize the ice making energy by using the three-point interpolation method and various temperature conditions. The final method was a simple calculation method based on the amount of energy required to process the ice making load, and to power the auxiliary devices required to make and store ice. This method relies heavily on an assumed COP value of 1.8. The values in the table below are shown in comparison to the baseline energy of 574 kWh/yr.

Table 5.1 Ice Making Energy by Various Methods – French Door Unit #1				
Method		Ice making energy		
		kWh/kg	kWh/yr	% of baseline energy
1	Median position only	0.413	123.0	21.4 %
2	Triangular Interpolation (3.9 °C/-17.8 °C)	0.375	112.0	19.5 %
	Triangulation (M/M) (6.2/-18.8)	0.396	118.3	20.6 %
	Triangulation (C/C) (3.5/-22.3)	0.436	129.9	22.6 %
	Triangulation (C/W) (2.9/-17.1)	0.363	108.4	18.9 %
3	Calculation	0.303	90.2	15.7 %

Error Associated with Incomplete Ice Maker Cycling

This product has a variable speed compressor which greatly simplifies the recognition of energy associated with the ice maker. In this case, the compressor power maintains a steady part-load condition throughout its operation, and there is no compressor cycling. Since there is no compressor cycling, the analysis was conducted using a whole number of ice making cycles and there is no error associated with incomplete cycling in this manner.

There is, however, a certain level of error due to the first batch of ice harvested upon the initiation of ice production. Once initiated, the ice maker begins by harvesting a batch of cubes, but the energy used to cool and freeze water into this set of cubes was expended long before the ice maker was rendered operative. Therefore, the energy recorded for the first batch of ice cubes is merely that of the ejection heater. This should have a much smaller impact on the end result than the error associated with the units employing cycling compressors, particularly since the ejection heaters are such a large portion of the total energy used by the ice maker. However, this particular unit does not take many ice making cycles to fill the ice bin, therefore this aspect is significant here.

The heaters for this unit use a total of 36 kJ to harvest each batch of ice, which weighed approximately 69 grams. Therefore, in each measurement the energy recorded for the first batch of ice was only 0.149 kJ/kg. If the mass of the first batch of ice is subtracted from each data set along with the ejection heater power, the following results shown in Table 5.2 are obtained, which are slightly higher than those shown in Table 5.1.

Table 5.2 Energy Consumption Error Associated with First Batch French Door Unit #1				
		kWh/kg		
	Method	Measured Value	Value Excluding First Batch	Difference
1	Median position only	0.413	0.443	7.3 %
2	Triangular Interpolation	0.374	0.403	7.8 %

Although it is not difficult to correct for the first batch of ice in this case, it may not be possible for every unit because it may not always be possible to determine the number of cubes harvested in each cycle, etc. as was seen with the other test subjects. There is, however, a simple approach to avoid this type of error entirely at the onset of testing. If the water supplied to the cabinet is interrupted, the ice molds would be emptied prior to the beginning of the data collection period. Therefore, one could begin the test by opening the water supply and then rendering the ice maker operative. After rendering the ice maker operative, the ejection heaters would operate and then the solenoid valve would fill the molds with water for the first batch of ice. The test period would then begin at the first solenoid event and end after the last batch of ice is harvested, thereby consisting of a whole number of ice-making cycles.

6: Experimental Results for French Door Refrigerator Freezer #2

This unit is a 765 liter (27 cubic foot) French door, energy star rated model with through the door ice and water service. It has variable defrost control, electronic thermostat control, and a variable-speed compressor. The unit has a factory-installed ice maker which is located in the upper portion of the refrigerator compartment. It is connected to an external water supply and produces ice in the ceiling of the refrigerator compartment and stores it in a bin located in the door of the refrigerator compartment. This unit freezes ice using a set of small ducts inside the walls of the cabinets to bring cold air from the freezer compartment located beneath the refrigerator compartment. The compartment temperatures used for the calculations that follow are the average of the time averaged values reported from each thermocouple during the test duration.

Figure 6.1 French Door Unit #2

The minimum and maximum time between defrost for this unit are 8 and 96 hours of compressor run time, respectively. According to the method outlined in the current DOE energy test procedure, these values result in a calculated average time between defrost of 30 hours of compressor run time.

This unit also has an automatically controlled anti-sweat heater, which responds to the ambient humidity condition. The manufacturer provided the control algorithm for this feature, and we maintained a dew point below 8 °C for all tests in order to ensure that this feature would not consume electricity outside the scope of this set of tests.

6.1 French Door Refrigerator Freezer #2 with No Ice Production
6.1.1 Mid Setting Results
During the first test, the energy consumption was measured at with the thermostat set to the median setting for each compartment. The following results were obtained:

Steady state operation time: 13497 seconds = 03:44:57
Measured refrigerator compartment temperature: (3.7 ± 0.1) °C
Measured freezer compartment temperature: (-16.8 ± 0.1) °C
Energy expended during the test period: (237.2 ± 1.2) watt-hours

Alternatively, this can be expressed as an average steady state power of (63.28 ± 0.32) watts at the measured temperatures. In alternative units, (554.3 ± 2.8) kWh/yr.

Measurements during the defrost sequence yielded the following results:

Defrost time: 8701 seconds = 02:25:01
Energy expended during defrost period: (303.3 ± 1.5) watt-hours

Combining the results of these two portions of the test, assuming that the average time between defrost is 30 hours, we arrive at the total energy consumption of (576.3 ± 2.9) kWh/yr at the measured temperatures.

6.1.2 Cold Setting Results
The second set of tests were conducted with the thermostat for each compartment set to the coldest setting. During this test, the compressor did not cycle on and off, but did exhibit temperature and power oscillations due to the part load compressor speed and the periodic cooling load shifts between the refrigerator and freezer compartment. In order to ensure that the most repeatable result was captured, the test period used consisted of a whole number of these oscillations. The following results were obtained:

Steady state operation time: 13622 seconds = 03:47:02
Measured refrigerator compartment temperature: (-1.2 ± 0.1) °C
Measured freezer compartment temperature: (-20.0 ± 0.1) °C
Energy expended during the test period: (287.7 ± 1.4) watt-hours

Alternatively, this can be expressed as an average steady state power of (76.04 ± 0.38) watts at the measured temperatures, or (666.1 ± 3.3) kWh/yr.

Measurements during the defrost sequence yielded the following results:

Defrost time: 14400 seconds = 04:00:00 (no cycling condition)
Energy expended during defrost period: (450.9 ± 2.3) watt-hours

Incorporating the defrost energy results in an overall energy consumption value of (687.6 ± 3.4) kWh/yr at these temperatures.

6.1.3 Mixed Setting Results

The thermostat for this unit was programmed in order to optimize performance while simultaneously maintaining temperatures of 7.2 °C and -15.0 °C in the refrigerator and freezer, respectively. Therefore, a third data set was taken with the refrigerator set to the warmest setting and the freezer set to the coldest setting in order to appropriate data for the three point interpolation method. These settings resulted in the compressor operating at part load conditions, but with periodic cycling. Because of this, the data was reduced using a whole number of complete compressor cycles. The third data set resulted in the following values:

Steady state operation time: 33243 seconds = 03:44:57
Measured refrigerator compartment temperature: (7.2 ± 0.1) °C
Measured freezer compartment temperature: (-20.3 ± 0.1) °C
Energy expended during the test period: (640.4 ± 3.2) watt-hours

Alternatively, this can be expressed as an average steady state power of (69.35 ± 0.35) watts at the measured temperatures, or (607.5 ± 3.0) kWh/yr. Measurements during the defrost sequence yielded the following results:

Defrost time: 23818 seconds = 06:36:58
Energy expended during defrost period: (615.0 ± 3.1) watt-hours

Incorporating the defrost energy results in an overall energy consumption value of (630.3 ± 3.2) kWh/yr at these temperatures.

6.1.4 Combined Test Results for French Door Refrigerator Freezer #2 with No Ice Production

The steady operation values from these three data sets were used to calculate the energy consumption at -17.8 °C and 3.9 °C using the three point interpolation method which resulted in an energy consumption value of (576 ± 3) kWh/yr. The total energy consumption, including the defrost energy, was (598 ± 3) kWh/yr.

6.2 French Door Refrigerator Freezer #2 with Ice Production

With the baseline energy consumption characterized, we next performed a series of measurements to examine the energy consumption attributed to the production of ice by the automatic ice maker.

<u>6.2.1 Mid Setting Results</u>
The thermostat was set to the median position for each compartment and the ice maker was rendered operative. Water was supplied to the unit at a temperature of 32.2 °C. Figure 6.2 shows the power signature of this unit while making ice under these conditions.

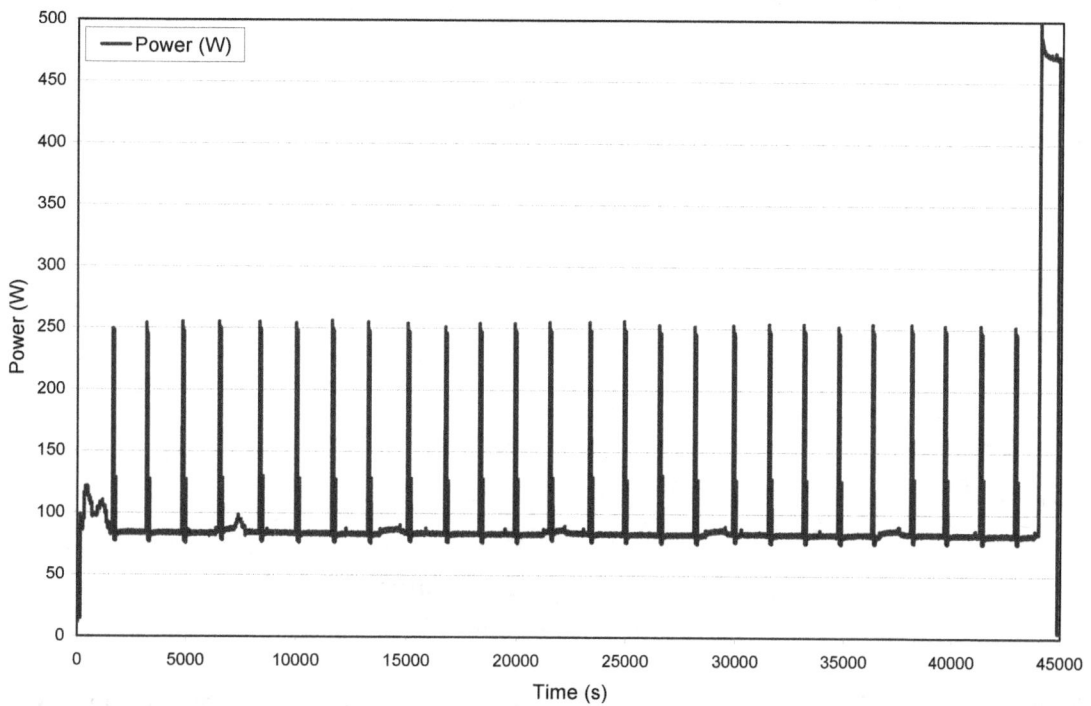

Figure 6.2 Refrigerator Power for Ice Production at Mid/Mid Setting for
French Door Unit #2

Starting from an empty ice bin condition, the ice maker operated for less than 44 000 seconds before undergoing a defrost period. This unit continued to produce ice during the defrost and recovery period, however, only the data leading up to the defrost was used for this analysis in order to separate out the effects of the defrost cycle. In total, 2.37 kg of ice was produced during 40 ice making cycles; 26 of these cycles occurred before the defrost; therefore the mass of 1.54 kg was used for this analysis. Each time the heater is powered on, it consumes 165 watts for approximately 70 seconds, and the solenoid consumes 50 watts for 4 seconds.

It is also noted that the freezer compartment became slightly warmer for a short period near the beginning of the data set, when the ice maker was switched on, and gradually decreased over the duration of the test period. This temperature increase was in response to the additional load realized by the freezer compartment as cold air was bled off to produce ice in the ice maker. It is also interesting to note that the temperature in the refrigerator compartment became slightly colder when the ice maker was rendered operative. This is because the ice maker is located in the refrigerator compartment and the additional supply of cold air needed to freeze ice has a cooling effect on the entire compartment.

49

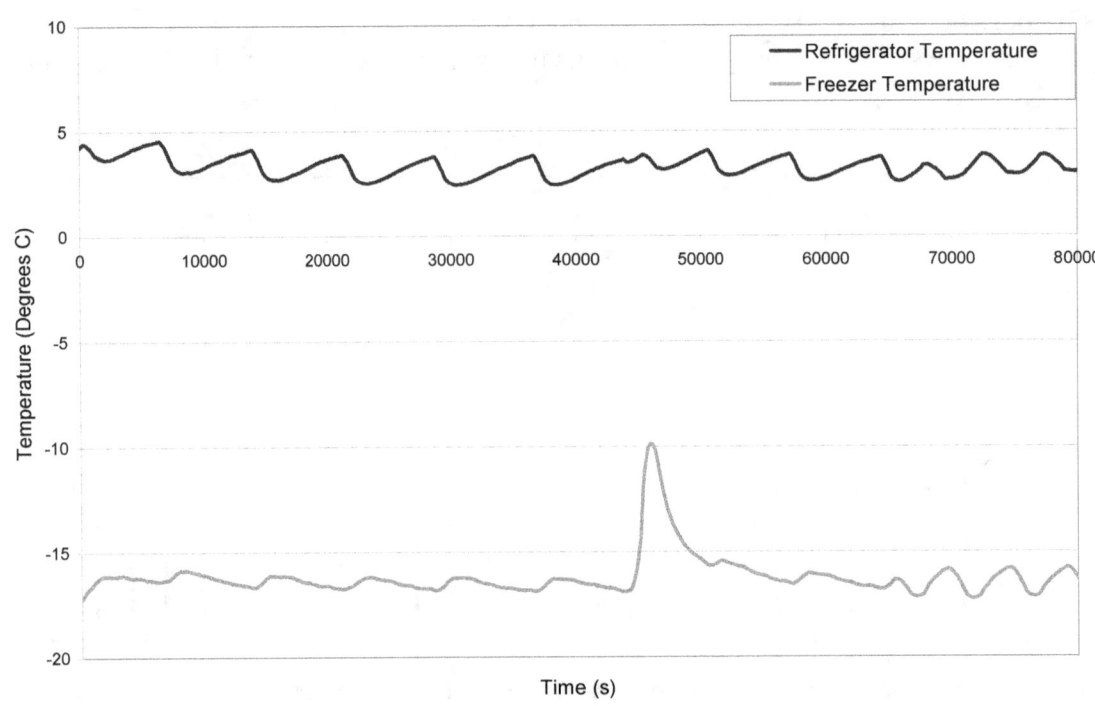

Figure 6.3 Temperature Response to Ice Production at Mid/Mid Setting for
French Door Unit #2

Since the compressor maintained a part load condition throughout the data set and did not switch on and off, it was not necessary to consider compressor cycles and the ice-making energy was calculated by considering all of the ice making cycles prior to the defrost. The test period is then described as starting from the ice maker being cued to produce ice until the moment that the defrost heater was activated. This is clearly visible on Figure 6.3.

Operation time: 41273 seconds = 11:27:53
Measured refrigerator compartment temperature: (3.2 ± 0.1) °C
Measured freezer compartment temperature: (-16.8 ± 0.1) °C
Mass of ice produced: (1.54 ± 0.005) kg
Ice production rate: (134.37 ± 0.44) grams per hour
Energy expended during the test period: (1052.0 ± 5.3) watt-hours

Alternatively, this can be expressed as an average steady state power of
(91.79 ± 0.46) watts at the measured temperatures; or (804.1 ± 4.0) kWh/yr.

Single Data Point Method for Measuring Ice-Maker Energy Consumption
The first proposed approach to quantifying the ice making energy is to compare the steady state energy consumption with both thermostats set to their median setting for both ice producing and ice maker inoperative conditions. In this case, the differential power consumption was:

50

$$(91.79 \pm 0.46) \text{ W} - (63.28 \pm 0.32) \text{ W} = (28.51 \pm 0.56) \text{ W}$$

The energy used to produce ice in this configuration can then be calculated from the mass of ice, the differential power and the test period time.

$$\frac{(0.02851 \text{ kW})(11.46 \text{ h})}{(1.54 \text{ kg})} = (0.212 \pm 0.004) \text{ kWh/kg}$$

Comparing the temperature results of the tests at the median setting shows that this particular unit maintained fairly consistent compartment temperatures whether or not the unit was producing ice. Although there was some transient behavior at the onset of ice making, the time averaged values showed that the freezer compartment temperature remained nearly identical temperatures, but the refrigerator compartment temperature was approximately 0.5 °C colder while making ice.

6.2.2 Cold Setting Results

The next approach was to attempt to use an interpolated energy (and ice production rate) using three sets of measurements. We set the thermostats of each compartment to the coldest setting for the second ice making data set. Figure 6.4 shows the power signature recorded during this measurement. This figure shows that the compressor power fluctuates under these conditions; this is because this unit constantly adjusts the compressor speed in order to maintain constant temperatures in the compartments. We averaged the data over the entire data set in order to dampen out these effects.

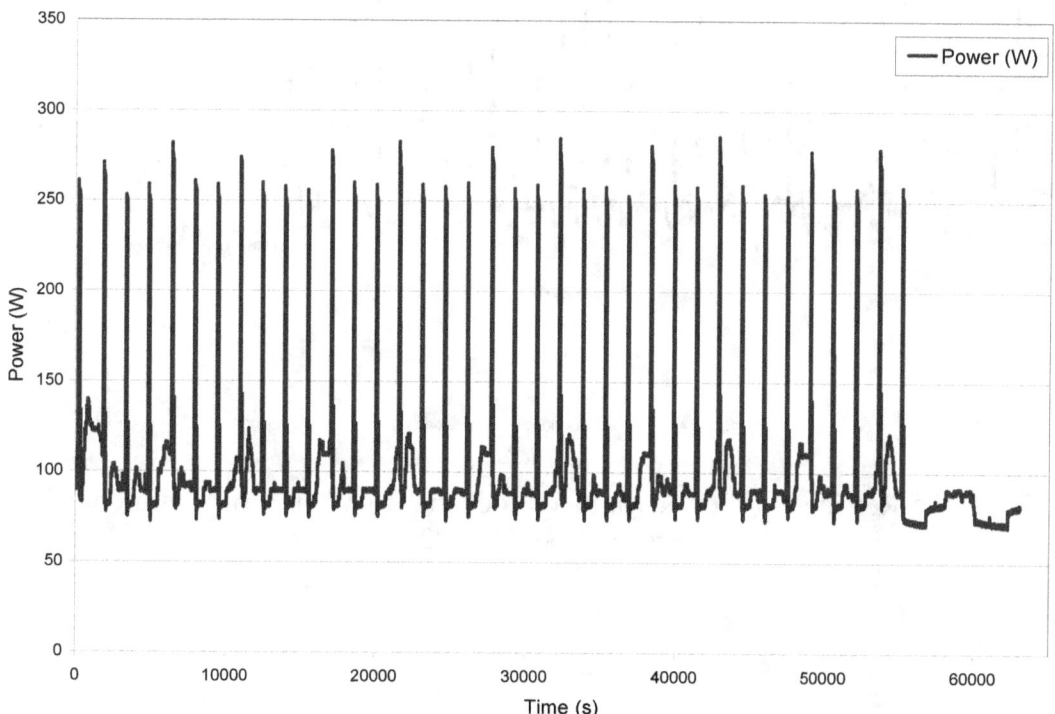

Figure 6.4 Power for Ice Production at Cold/Cold Setting
French Door Unit #2

51

Operation time: 54928 seconds = 15:15:28
Measured refrigerator compartment temperature: (-0.1 ± 0.1) °C
Measured freezer compartment temperature: (-18.2 ± 0.1) °C
Mass of ice produced: (1.800 ± 0.005) kg
Ice production rate: (117.97 ± 0.33) grams per hour
Energy expended during the test period: (1541.9 ± 7.7) watt-hours

Alternatively, this can be expressed as an average steady state power of
(101.06 ± 0.51) watts at the measured temperatures; or (885.3 ± 4.4) kWh/yr.

6.2.3 Mixed Setting Results

The final measurement for this unit was performed with the freezer set to the coldest
setting and the refrigerator set to the warmest setting. This data set showed a gradual
change in temperatures during the initiation and termination of ice production. It also
displayed an abrupt reduction in compressor speed and a return to periodic oscillation at
the time that it ceased ice production, shown in Figure 6.5.

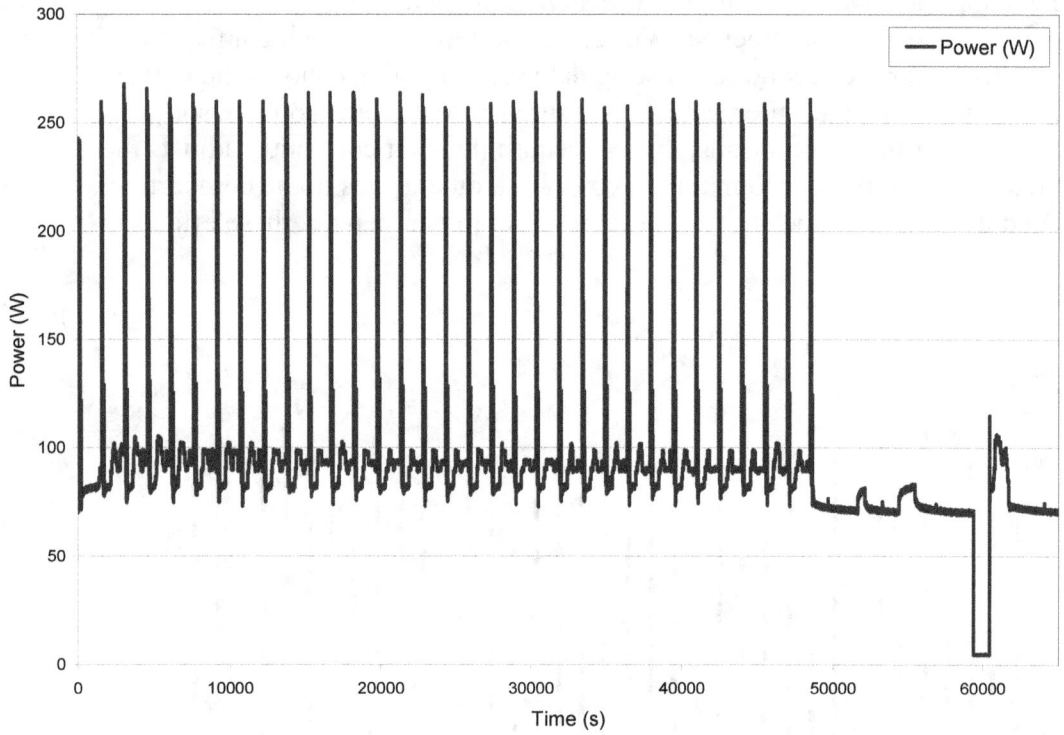

Figure 6.5 Refrigerator Power with Ice Production at Warm/Cold Setting for
French Door Unit #2

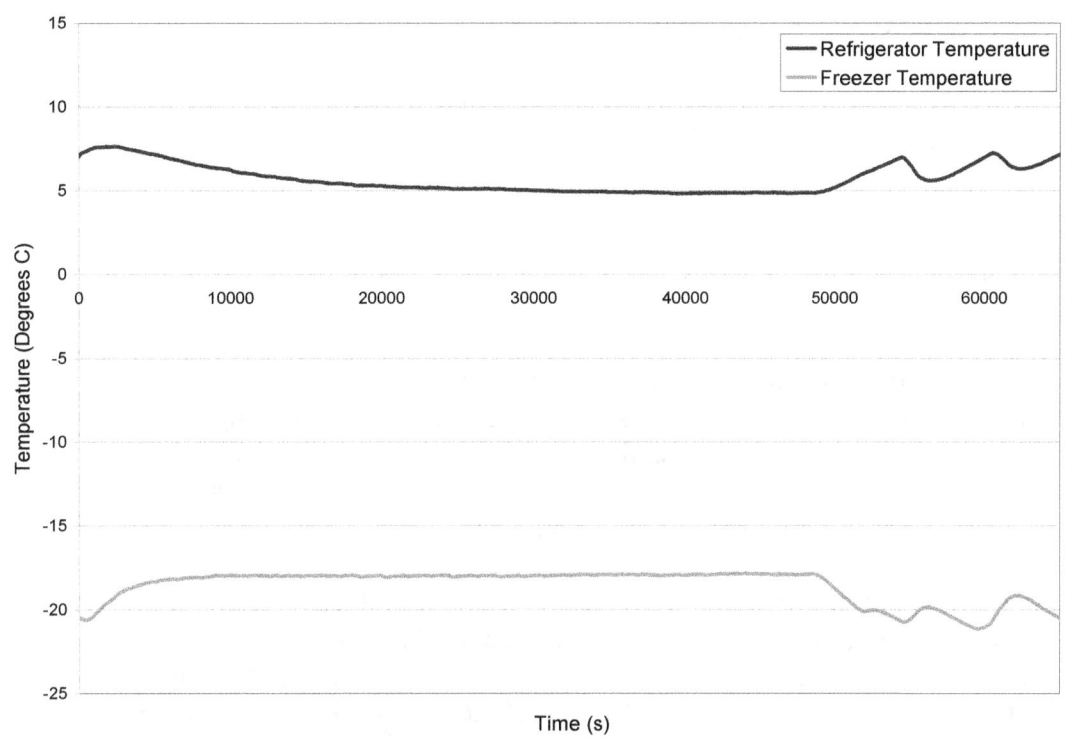

Figure 6.6 Refrigerator and Freezer Compartment Temperatures with Ice Production at Warm/Cold Setting for French Door Unit #2

The following data were obtained from these measurements:

Operation time: 48398 seconds = 13:26:38
Measured refrigerator compartment temperature: (5.4 ± 0.1) °C
Measured freezer compartment temperature: (-18.4 ± 0.1) °C
Mass of ice produced: (1.57 ± 0.005) kg
Ice production rate: (116.9 ± 0.4) grams per hour
Energy expended during the test period: (1324.1 ± 6.6) watt-hours

Alternatively, this can be expressed as an average steady state power of (98.5 ± 0.5) watts at the measured temperatures; or (862.8 ± 4.3) kWh/yr.

Three Data Point Method for Measuring Ice-Maker Energy Consumption
By applying the three point interpolation method for both the energy consumption and the ice production rate, the following values were obtained for conditions representative of 3.9 °C in the refrigerator and -17.8 °C in the freezer.

Energy consumption: (845 ± 4) kWh/yr ((96.4 ± 0.5) W)
Ice production rate: (2.96 ± 0.02) kg/day ((123.3 ± 0.9) grams per hour)

The following results were obtained by comparing the triangularly interpolated results from the tests with and without ice production.

53

$$(96.4 \pm 0.5) \text{ W} - (65.8 \pm 0.3) \text{ W} = (30.6 \pm 0.6) \text{ W}$$

The energy used to produce ice in this configuration can then be calculated from the mass of ice, the differential power and the test period time.

$$\frac{(0.0306 \text{ kW})}{\left(0.1233 \frac{\text{kg}}{\text{h}}\right)} = (0.248 \pm 0.005) \text{ kWh/kg}$$

Next, it is of interest to calculate the ice-making energy by triangular interpolation of the non ice-making data to the temperatures realized during the ice-making tests. First, the ice making test data at the mid/mid setting resulted in temperatures of 3.2 °C in the refrigerator and -16.8 °C in the freezer. Triangularly interpolating the non-ice making data points to these temperatures, we calculate a value of (63.7 ± 0.5) W, which is (28.1 ± 0.7) W less than the average power seen during the mid/mid test with ice production. Dividing this differential power by the ice production rate, the ice making energy at T_{fridge}=3.2 °C and T_{freeze}=-16.8 °C is (0.209 ± 0.005) kWh/kg. Similarly, the ice-making energy was calculated at the coldest and mixed settings as well. The coldest setting, at T_{fridge}=-0.1 °C and T_{freeze}=-18.2 °C is (0.260 ± 0.009) kWh/kg; the mixed setting, at T_{fridge}=5.4 °C and T_{freeze}=-18.4 °C is (0.278 ± 0.005) kWh/kg. These results confirm the observation from the previous unit that the ice-making energy is fairly dependent on the refrigerator and freezer compartment temperatures.

6.3 Calculation Based Method

Lastly, the theoretical ice-making energy was calculated using the simplified calculation approach. Each data set taken with the ice maker operative was started with the ice bin empty and logged until the bin reached its shut off limit. The first data set showed 1.54 kg of ice produced through 26 cycles, the second showed 1.8 kg in 36 cycles, and the third showed 1.57 kg in 32 cycles. On average, this unit produces 52.2 grams of ice during each cycle.

The solenoid valve draws 50 watts for a period of 4 seconds to deliver water to the molds which results in 200 Joules, or 3.83 kJ/kg of ice produced. The ice ejection heater draws 165 watts for 70 second periods to free the ice from the molds which results in 11,550 Joules, or 221.26 kJ/kg of ice produced. The total energy consumed by these components is therefore 225.10 kJ/kg of ice produced. Also, the energy consumed by the ice ejection heater is ultimately dissipated as heat into the ice-making compartment; therefore it must be considered an additional heat load to be processed by the refrigeration system.

Considering the energy required to cool 32.2 °C water to freezing and the additional load of the heaters, the refrigeration system must therefore meet the additional demand of to remove 726 kJ of heat per kg of ice produced. Using the assumed COP of 1.8, the additional compressor duty to remove this heat will total 403.4 kJ/kg. Adding in the energy consumed by the solenoid and ejection heater brings the total load to 628.5 kJ/kg; or in alternative units 0.175 kWh/kg.

6.4 Summary

The ice making energy was calculated by 4 methods as shown in Table 6.1. the table also shows the energy impact of producing 0.816 kg of ice per day.

The first method was to compare the difference in energy consumed with the thermostats set to their median positions. This method resulted in a temporary departure from the refrigerator cabinet temperature because of the control system used in this cabinet. The second method attempts to characterize the ice making energy by using the various three-point interpolation methods. The final method was a simple calculation method based on the amount of energy required to process the ice making load, and to power the auxiliary devices required to make and store ice. The values in Table 6.1 are shown in comparison to the baseline energy of 598 kWh/yr.

Table 6.1 Ice Making Energy by Various Methods – French Door Unit #2				
Method		Ice making energy		
		kWh/kg	kWh/yr	% of baseline energy
1	Median position only	0.212	63.1	10.6 %
2	Triangulation (3.9/-17.8)	0.248	74.1	12.4 %
	Triangulation (M/M) (3.2/-16.8)	0.209	62.2	10.4 %
	Triangulation (C/C) (-0.1/-18.2)	0.260	78.0	13.0 %
	Triangulation (W/C) (5.4/-18.4)	0.278	82.8	13.8 %
3	Calculation	0.175	52.1	8.72 %

Error Associated with Incomplete Ice-Maker Cycling

This product, like the other French door unit, has a variable-speed compressor which simplified the recognition of energy associated with the ice maker because the compressor never cycled off during ice production. Since there was no compressor cycling during the ice-making tests, the analysis was conducted using a whole number of ice-making cycles and there is no error associated with incomplete cycling.

As was the case with the French Door Refrigerator Freezer Unit #1, the level of error due to the first batch of ice harvested upon the initiation of ice production is examined in this section. Performing a similar analysis yields significant results. As mentioned in the previous section, however, the influence of this first batch of ice on the energy consumption can be easily eliminated by using the valve approach described in the previous section of this report.

Table 6.2 Energy Consumption Error Associated with First Batch French Door Unit #2				
		kWh/kg		
	Method	Measured Value	Value Excluding First Batch	Difference
1	Median position only	0.212	0.219	3.3 %
2	Triangular Interpolation	0.248	0.257	3.6 %

7: Observations

Four refrigerators with automatic ice makers were examined in this study to evaluate test procedure options for determining energy consumption due to ice production by automatic ice makers. The units varied in configuration, technological sophistication, and most importantly ice-making energy consumption. This study centered on an approach that was based on using the longest attainable period of test data that included steady state or cyclic operation. The length of the test period for each data set was limited by either a complete filling of the ice storage bin or by an interruption of steady operation due to a defrost event.

The first two units analyzed employed a single-speed compressor that maintained temperature by cycling on and off. For these units, the ice-making cycles did not typically coincide with the compressor cycles, which compromised the reproducibility of the test data. However, the reproducibility of the end result is strictly a function of the number of undisturbed sequential cycles that can be found within a data set. A unit with a very large ice storage bin operating in such a way that it shows a long time between defrost events will be simpler to evaluate than one with a small storage bin or a short time between defrost events.

The second two units studied were French door units. These units were particularly interesting because they produce ice inside the refrigerator compartment rather than the freezer compartment, which was expected to result in less efficient ice production. However, these are high-end units that employ sophisticated features. They both use inverter-driven variable-speed compressors, which are more efficient than single-speed compressors. These units ultimately showed better ice-making energy consumption values than the units that made ice in the freezer. One particularly interesting aspect of the units with variable-speed compressors is that the ice-making cycles can be easily separated from the data because the compressors do not cycle on and off. This greatly simplifies the determination of ice making energy.

This study also considered a purely calculation-based approach to estimate the ice-maker energy consumption. This method proved to be undesirable because the ice making energy consumption must be based on the actual mass of ice produced, which must be a measured quantity. Also, this approach is based on an assumed COP of the system, which is not a traceable or reliable quantity.

The data showed that the compartment temperatures typically deviate from their set points when the ice maker is rendered operative. Furthermore, the ice-making energy consumption generally varies with the compartment temperatures. For this reason, it is strongly recommended that the ice-making energy be determined at a specified standard set of cabinet temperatures in order to have a good basis for comparison of products. It is also recommended that a three point triangular interpolation method be used to determine the ice-making energy consumption if two point interpolation does not closely approximate the target refrigerator and freezer temperatures simultaneously.

The most reliable and cross-comparative results were those obtained using the interpolation methods and the target temperatures of 3.9 °C in the refrigerator and -17.8 °C in the freezer. Examination of these results shows that the measured ice maker energy consumption ranges from 0. 652 kWh/kg down to 0.249 kWh/kg.   In general, roughly half of the energy expended during the ice production process was due to the operation of electric resistance heaters, and it is estimated that approximately half of the remaining energy was expended to remove the heat added by these heaters. In other words, approximately only one quarter of the total energy used to make ice is actually used to cool and freeze water into ice. This indicates that there are substantial opportunities for efficiency improvements merely by optimizing the operation of the heaters associated with the ice makers or by eliminating them altogether and using an alternative method to free the ice from the molds.

It should also be noted that the theoretical limit of ice-making energy is merely the difference between the inlet water's enthalpy and that of the ice in its final state, divided by the vapor compression system's COP. The enthalpy difference is a firm parameter of 505 kJ/kg for water entering the ice maker at 32.2 °C (90 °F) and being brought to a final temperature of -17.8 °C (0 °F). The COP is limited by the Carnot efficiency of a vapor compression cycle operating between a condensing temperature of 32.2 °C and an evaporating temperature of -17.8 °C, at these conditions COP_{Carnot} is 5.1. Therefore, the theoretical limit of ice making energy consumption is 99 kJ/kg or 0.028 kWh/kg. Although this is an unattainable limit in practice, it provides insight to the fact that the best technology evaluated in this study is an order of magnitude less efficient than the theoretical limit.

8: Recommendations for Ice-Making Energy Consumption Test

Through this study, several key aspects of a repeatable, reproducible ice-making energy consumption test have been determined. In order to minimize overall test burden, the several parameters of an ice making energy consumption test can be aligned with the basic energy consumption test for domestic refrigerating appliances including ambient temperature in test chamber of 32.2 °C (90 °F) and water inlet temperature of 32.2 °C (90 °F), since it is not practical to deliver water to the test unit at a temperature different from the ambient condition.

The ice-making energy consumption should be determined by measuring the energy consumption of the unit while actively producing ice and subtracting the energy consumption of the unit while not actively producing ice. Each of these quantities should be determined by interpolating the results of individual measurements to a set of compartment temperatures; specifically 3.9 °C in the refrigerator compartment and -17.8 °C in the freezer compartment to maintain alignment with the energy consumption test. The ice production rate should be interpolated to these temperatures by the same means.

The result of the ice-making energy consumption measurement is then defined as the ice-making energy consumption at the specified target temperatures divided by the ice production rate at the specified target temperatures.

The test period for the ice making energy consumption test must include enough data to ensure a repeatable and reproducible measurement. This is a difficult parameter to define for units that use a single-speed compressor that maintains conditions in the compartments by switching on and off. The recommendation for this type of unit is to perform the energy consumption test over a whole number of compressor cycles and to cumulatively calculate the ratio of cumulative average of the power draw to the total mass of ice produced through all of these compressor cycles. The data set is considered sufficiently long if this parameter does not change significantly when another compressor cycle is included in the analysis.

If a defrost event interrupts the data set, only the portion of the data collected during the steady operation period should be used. The mass of ice produced during the steady portion may be estimated by multiplying the total mass of ice produced by the proportion of ice making cycles that occurred during this period. If the longest attainable data set does not include enough data to satisfy the steady state criteria mentioned above, the test may be repeated and the usable cycles from that data set may be assembled with those from the previous test(s) to satisfy the criteria.

For the case of a variable-speed compressor, the test is greatly simplified because the test period is not affected by compressor cycles. In this case, the water supply line should be outfitted with a valve that can stop the supply of water to the cabinet. With the valve shut, the ice maker can be operated until the ice molds are emptied. The ice-making energy consumption test can begin by opening the water supply valve and simultaneously rendering the ice maker operative. The test cabinet will then begin operation of the ice

59

maker with an empty bin condition, and the test period can be taken from the first solenoid opening to the last ice ejection heater operation; a whole number of ice-making cycles.

9: Open Items for Future Consideration

Through this study, several issues became apparent for consideration of future investigation. The first step is to validate the recommended methodology through additional testing. We will consider the developed measurement techniques as an adequate tool for rating these products if the results of multiple measurements of several automatic ice makers can be reproduced at different laboratories.

Another issue is that the measurement techniques must be validated for use with other styles of automatic ice makers. Four styles of automatic ice makers were examined in the present study, the energy use of each was dominated by electric resistance heaters used to free frozen ice from the ice maker. Other technologies currently exist for ejecting ice from the ice makers and it is of interest to examine test data for such units to ensure that such a method is applicable for these technologies.

The objective of an automatic ice maker is to produce a quantity of ice, not maintain a specific condition; therefore, it is not appropriate to measure the amount of energy that an automatic ice maker uses during a specified amount of time. Instead, such products must be rated on their ice-making energy consumption, which is the amount of energy expended normalized to the amount of ice produced. However, the value of ice making energy consumption can be scaled to an appropriate annualized energy consumption using a usage factor that represents the amount of ice that the units actually produce in a typical residential setting. To date, there is limited information regarding consumer usage of automatic ice makers and it is recommended that studies be conducted to seek this information.

Throughout this study, the effects of ice making on the contribution of energy related to periodic defrosting was not considered. Operation of automatic ice makers may, under certain conditions, strongly impact the defrost frequency of certain refrigerator designs. This is a secondary issue but is very complex and would be difficult to factor into a rated energy consumption value for an ice maker; however, the impact may be quite substantial as evidenced by the first unit examined in this study.

Finally, this study relied on an assumption that the ice production rate at the specific target temperatures could be estimated by interpolating the results of the measurements in the same manner as energy consumption. Although this method is a good first approximation, it is necessary to either verify this method or to quantify the error associate with this approach.

10: References

AHAM, 1979. ANSI/AHAM HRF-1-1979, American National Standard for Household Refrigerators and Household Freezers. Chicago: Association of Home Appliance Manufacturers.

AHAM, 2009. AHAM Update to DOE on Status of Ice Maker Energy Test Procedure. November 19, 2009.

ASNZ, 2007. ASNZ 4474.1 Performance of household electrical appliances— Refrigerating appliances. Part 1 Energy Consumption and Performance.

Uniform Test Method for Measuring the Energy Consumption of Electric Refrigerators and Electric Refrigerator-Freezers, 10 Federal Register 430, Subpart B, Appendix A1 (01JAN2010), pp. 159-167.

Haider, I., Feng, H., and Radermacher, R., 1996. Experimental Results of a Household Automatic Icemaker in a Refrigerator/Freezer. ASHRAE Transactions: Symposia, SA-96-7-3, pp. 541-545.

Meier, A. and Martinez, M., 1996. Energy Use of Ice Making in Domestic Refrigerators. ASHRAE Transactions, Vol., 102, Pr. 2, pp. 1071-1076.

Appendix: Uncertainty Analysis

The equation used to calculate the energy consumption using 2 point interpolation is as follows:

$$E = \frac{E_1\left(T_{ref} - T_2\right) - E_2\left(T_{ref} - T_1\right)}{T_1 - T_2}$$

Where E is the energy consumption (kWh/yr) and T is the compartment temperature (°C), the subscripts 1 and 2 refer to the data point and 'ref' refers to the standardized reference temperature at which the equation calculates the energy consumption.

In order to calculate the propagation of measurement uncertainty, it is necessary to take the partial derivatives of energy with respect to each measured quantity:

$$\frac{\partial E}{\partial T_1} = \frac{\left(E_1 - E_2\right)\left(T_{ref} - T_2\right)}{\left(T_1 - T_2\right)^2}$$

$$\frac{\partial E}{\partial E_1} = \frac{\left(T_{ref} - T_2\right)}{\left(T_1 - T_2\right)}$$

$$\frac{\partial E}{\partial T_2} = \frac{\left(E_1 - E_2\right)\left(T_{ref} - T_1\right)}{\left(T_1 - T_2\right)^2}$$

$$\frac{\partial E}{\partial E_2} = \frac{\left(T_1 - T_{ref}\right)}{\left(T_1 - T_2\right)}$$

And the uncertainty of the calculated energy consumption is:

$$U_E = \sqrt{\left(\frac{\partial E}{\partial T_1} U_{T_1}\right)^2 + \left(\frac{\partial E}{\partial E_1} U_{E_1}\right)^2 + \left(\frac{\partial E}{\partial T_2} U_{T_2}\right)^2 + \left(\frac{\partial E}{\partial E_2} U_{E_2}\right)^2}$$

Where U is the uncertainty of the value specified by the accompanying subscript, in the same units as the quantity.

The uncertainty analysis for the triangular interpolation approach is much more complicated since; it incorporates the measurements from both compartments simultaneously. The subscripts 'A' and 'B' are used to denote Refrigerator and Freezer compartments. The first step is to calculate the location of a theoretical fourth point from the following relationship:

$$T_{A4} = \left(T_{xB} - \frac{T_{xA}(T_{B2} - T_{xB})}{(T_{A2} - T_{xA})} - T_{B1} + \frac{T_{A1}(T_{B3} - T_{B1})}{(T_{A3} - T_{A1})} \right) \left(\frac{(T_{B3} - T_{B1})}{(T_{A3} - T_{A1})} - \frac{(T_{B2} - T_{xB})}{(T_{A2} - T_{xA})} \right)^{-1}$$

Or alternatively written:

$$T_{A4} = \frac{T_{A1}T_{A2}T_{B3} - T_{A1}T_{A2}T_{xB} + T_{A1}T_{xA}T_{B2} - T_{A1}T_{xA}T_{B3} - T_{A2}T_{A3}T_{B1} + T_{A2}T_{A3}T_{xB} + T_{A3}T_{xA}T_{B1} - T_{A3}T_{xA}T_{B2}}{T_{A1}T_{B2} - T_{A1}T_{xB} - T_{A2}T_{B1} + T_{A2}T_{B3} - T_{A3}T_{B2} + T_{A3}T_{xB} + T_{xA}T_{B1} - T_{xA}T_{B3}}$$

The partial derivatives of T_{A4} with respect to each of the other measured temperatures are:

$$\frac{\partial T_{A4}}{\partial T_{A1}} = \frac{(T_{A2}T_{B3} - T_{A2}T_{xB} + T_{xA}T_{B2} - T_{xA}T_{B3})}{(T_{A1}T_{B2} - T_{A1}T_{xB} - T_{A2}T_{B1} + T_{A2}T_{B3} - T_{A3}T_{B2} + T_{A3}T_{xB} + T_{xA}T_{B1} - T_{xA}T_{B3})}$$

$$\frac{-(T_{B2} - T_{xB})(T_{A1}T_{A2}T_{B3} - T_{A1}T_{A2}T_{xB} + T_{A1}T_{xA}T_{B2} - T_{A1}T_{xA}T_{B3} - T_{A2}T_{A3}T_{B1} + T_{A2}T_{A3}T_{xB} + T_{A3}T_{xA}T_{B1} - T_{A3}T_{xA}T_{B2})}{(T_{A1}T_{B2} - T_{A1}T_{xB} - T_{A2}T_{B1} + T_{A2}T_{B3} - T_{A3}T_{B2} + T_{A3}T_{xB} + T_{xA}T_{B1} - T_{xA}T_{B3})^2}$$

$$\frac{\partial T_{A4}}{\partial T_{A2}} = \frac{(T_{A1}T_{B3} - T_{A1}T_{xB} - T_{A3}T_{B1} + T_{A3}T_{xB})}{(T_{A1}T_{B2} - T_{A1}T_{xB} - T_{A2}T_{B1} + T_{A2}T_{B3} - T_{A3}T_{B2} + T_{A3}T_{xB} + T_{xA}T_{B1} - T_{xA}T_{B3})}$$

$$\frac{-(-T_{B1} + T_{B3})(T_{A1}T_{A2}T_{B3} - T_{A1}T_{A2}T_{xB} + T_{A1}T_{xA}T_{B2} - T_{A1}T_{xA}T_{B3} - T_{A2}T_{A3}T_{B1} + T_{A2}T_{A3}T_{xB} + T_{A3}T_{xA}T_{B1} - T_{A3}T_{xA}T_{B2})}{(T_{A1}T_{B2} - T_{A1}T_{xB} - T_{A2}T_{B1} + T_{A2}T_{B3} - T_{A3}T_{B2} + T_{A3}T_{xB} + T_{xA}T_{B1} - T_{xA}T_{B3})^2}$$

$$\frac{\partial T_{A4}}{\partial T_{A3}} = \frac{(-T_{A2}T_{B1} + T_{A2}T_{xB} + T_{xA}T_{B1} - T_{xA}T_{B2})}{(T_{A1}T_{B2} - T_{A1}T_{xB} - T_{A2}T_{B1} + T_{A2}T_{B3} - T_{A3}T_{B2} + T_{A3}T_{xB} + T_{xA}T_{B1} - T_{xA}T_{B3})}$$

$$\frac{-(-T_{B2} + T_{xB})(T_{A1}T_{A2}T_{B3} - T_{A1}T_{A2}T_{xB} + T_{A1}T_{xA}T_{B2} - T_{A1}T_{xA}T_{B3} - T_{A2}T_{A3}T_{B1} + T_{A2}T_{A3}T_{xB} + T_{A3}T_{xA}T_{B1} - T_{A3}T_{xA}T_{B2})}{(T_{A1}T_{B2} - T_{A1}T_{xB} - T_{A2}T_{B1} + T_{A2}T_{B3} - T_{A3}T_{B2} + T_{A3}T_{xB} + T_{xA}T_{B1} - T_{xA}T_{B3})^2}$$

$$\frac{\partial T_{A4}}{\partial T_{B1}} = \frac{(-T_{A2}T_{A3} + T_{A3}T_{xA})}{(T_{A1}T_{B2} - T_{A1}T_{xB} - T_{A2}T_{B1} + T_{A2}T_{B3} - T_{A3}T_{B2} + T_{A3}T_{xB} + T_{xA}T_{B1} - T_{xA}T_{B3})}$$

$$\frac{-(-T_{A2} + T_{xA})(T_{A1}T_{A2}T_{B3} - T_{A1}T_{A2}T_{xB} + T_{A1}T_{xA}T_{B2} - T_{A1}T_{xA}T_{B3} - T_{A2}T_{A3}T_{B1} + T_{A2}T_{A3}T_{xB} + T_{A3}T_{xA}T_{B1} - T_{A3}T_{xA}T_{B2})}{(T_{A1}T_{B2} - T_{A1}T_{xB} - T_{A2}T_{B1} + T_{A2}T_{B3} - T_{A3}T_{B2} + T_{A3}T_{xB} + T_{xA}T_{B1} - T_{xA}T_{B3})^2}$$

$$\frac{\partial T_{A4}}{\partial T_{B2}} = \frac{\left(T_{A1}T_{xA} - T_{A3}T_{xA}\right)}{\left(T_{A1}T_{B2} - T_{A1}T_{xB} - T_{A2}T_{B1} + T_{A2}T_{B3} - T_{A3}T_{B2} + T_{A3}T_{xB} + T_{xA}T_{B1} - T_{xA}T_{B3}\right)}$$

$$-\frac{\left(T_{A1} - T_{A3}\right)\left(T_{A1}T_{A2}T_{B3} - T_{A1}T_{A2}T_{xB} + T_{A1}T_{xA}T_{B2} - T_{A1}T_{xA}T_{B3} - T_{A2}T_{A3}T_{B1} + T_{A2}T_{A3}T_{xB} + T_{A3}T_{xA}T_{B1} - T_{A3}T_{xA}T_{B2}\right)}{\left(T_{A1}T_{B2} - T_{A1}T_{xB} - T_{A2}T_{B1} + T_{A2}T_{B3} - T_{A3}T_{B2} + T_{A3}T_{xB} + T_{xA}T_{B1} - T_{xA}T_{B3}\right)^2}$$

$$\frac{\partial T_{A4}}{\partial T_{B3}} = \frac{\left(T_{A1}T_{A2} - T_{A1}T_{xA}\right)}{\left(T_{A1}T_{B2} - T_{A1}T_{xB} - T_{A2}T_{B1} + T_{A2}T_{B3} - T_{A3}T_{B2} + T_{A3}T_{xB} + T_{xA}T_{B1} - T_{xA}T_{B3}\right)}$$

$$-\frac{\left(T_{A2} - T_{xA}\right)\left(T_{A1}T_{A2}T_{B3} - T_{A1}T_{A2}T_{xB} + T_{A1}T_{xA}T_{B2} - T_{A1}T_{xA}T_{B3} - T_{A2}T_{A3}T_{B1} + T_{A2}T_{A3}T_{xB} + T_{A3}T_{xA}T_{B1} - T_{A3}T_{xA}T_{B2}\right)}{\left(T_{A1}T_{B2} - T_{A1}T_{xB} - T_{A2}T_{B1} + T_{A2}T_{B3} - T_{A3}T_{B2} + T_{A3}T_{xB} + T_{xA}T_{B1} - T_{xA}T_{B3}\right)^2}$$

The uncertainty of the temperature in compartment A is:

$$U_{T_{A4}} = \sqrt{\left(\frac{\partial T_{A4}}{\partial T_{A1}}U_{A_1}\right)^2 + \left(\frac{\partial T_{A4}}{\partial T_{A2}}U_{A2}\right)^2 + \left(\frac{\partial T_{A4}}{\partial T_{A3}}U_{A3}\right)^2 + \left(\frac{\partial T_{A4}}{\partial T_{B1}}U_{B_1}\right)^2 + \left(\frac{\partial T_{A4}}{\partial T_{B2}}U_{B2}\right)^2 + \left(\frac{\partial T_{A4}}{\partial T_{B3}}U_{B3}\right)^2}$$

Once the uncertainty of Temperature T_{A4} is defined, the uncertainty of the corresponding energy, E_4, can be calculated from the equation for E_4.

$$E_4 = E_1 + \left[\left(E_3 - E_1\right)\frac{\left(T_{A4} - T_{A1}\right)}{\left(T_{A3} - T_{A1}\right)}\right]$$

The partial derivatives of E_4 with respect to each measured quantity are:

$$\frac{\partial E_4}{\partial E_1} = 1 - \frac{\left(T_{A4} - T_{A1}\right)}{\left(T_{A3} - T_{A1}\right)}$$

$$\frac{\partial E_4}{\partial E_3} = \frac{\left(T_{A4} - T_{A1}\right)}{\left(T_{A3} - T_{A1}\right)}$$

$$\frac{\partial E_4}{\partial T_{A1}} = \left(E_3 - E_1\right)\left(\frac{-1}{\left(T_{A3} - T_{A1}\right)} + \frac{\left(T_{A4} - T_{A1}\right)}{\left(T_{A3} - T_{A1}\right)^2}\right)$$

$$\frac{\partial E_4}{\partial T_{A3}} = -\frac{\left(E_3 - E_1\right)\left(T_{A4} - T_{A1}\right)}{\left(T_{A3} - T_{A1}\right)^2}$$

$$\frac{\partial E_4}{\partial T_{A4}} = \frac{(E_3 - E_1)}{(T_{A3} - T_{A1})}$$

The uncertainty of E$_4$ is therefore:

$$U_{E_4} = \sqrt{\left(\frac{\partial E_4}{\partial E_1} U_{E_1}\right)^2 + \left(\frac{\partial E_4}{\partial E_3} U_{E_3}\right)^2 + \left(\frac{\partial E_4}{\partial T_{A1}} U_{T_{A1}}\right)^2 + \left(\frac{\partial E_4}{\partial T_{A3}} U_{T_{A3}}\right)^2 + \left(\frac{\partial E_4}{\partial T_{A4}} U_{T_{A4}}\right)^2}$$

Finally, the uncertainty of the overall energy consumption at the triangulated coordinates can be calculated from its equation:

$$E_x = E_2 + \left[(E_4 - E_2)\frac{(T_{xA} - T_{A2})}{(T_{A4} - T_{A2})}\right]$$

The partial derivatives of E$_x$ with respect to each measured quantity are:

$$\frac{\partial E_x}{\partial E_2} = 1 - \frac{(T_{xA} - T_{A2})}{(T_{A4} - T_{A2})}$$

$$\frac{\partial E_x}{\partial E_4} = \frac{(T_{xA} - T_{A2})}{(T_{A4} - T_{A2})}$$

$$\frac{\partial E_x}{\partial T_{A2}} = \frac{E_2 T_{A4} - E_2 T_{xA} - E_4 T_{A4} + E_4 T_{xA}}{(T_{A4} - T_{A2})^2}$$

$$\frac{\partial E_x}{\partial T_{A4}} = \frac{E_2 T_{xA} - E_2 T_{A2} + E_4 T_{A2} - E_4 T_{xA}}{(T_{A4} - T_{A2})^2}$$

The uncertainty of E$_x$ is therefore:

$$U_{E_x} = \sqrt{\left(\frac{\partial E_x}{\partial E_2} U_{E_2}\right)^2 + \left(\frac{\partial E_x}{\partial E_4} U_{E_4}\right)^2 + \left(\frac{\partial E_x}{\partial T_{A2}} U_{T_{A2}}\right)^2 + \left(\frac{\partial E_x}{\partial T_{A4}} U_{T_{A4}}\right)^2}$$